Mathematical Models of Social Evolution

Mathematical Models of Social Evolution:
A Guide for the Perplexed

Richard McElreath and Robert Boyd

The University of Chicago Press   Chicago and London

Richard McElreath is associate professor of anthropology and member of graduate groups in ecology, animal behavior and population biology, University of California, Davis.

Robert Boyd is professor of anthropology at the University of California, Los Angeles.

The University of Chicago Press, Chicago 60637
The University of Chicago Press, Ltd., London
© 2007 by The University of Chicago
All rights reserved. Published 2007
Printed in the United States of America

16 15 14 13 12 11 10 09 08 07   1 2 3 4 5

ISBN-13: 978-0-226-55826-4 (cloth)
ISBN-13: 978-0-226-55827-1 (paper)
ISBN-10: 0-226-55826-6 (cloth)
ISBN-10: 0-226-55827-4 (paper)

Library of Congress Cataloging-in-Publication Data

McElreath, Richard, 1973–
    Mathematical models of social evolution/Richard McElreath and Robert Boyd.
        p. cm.
    Includes bibliographical references and index.
    ISBN-13: 978-0-226-55826-4 (cloth : alk. paper)
    ISBN-10: 0-226-55826-6 (cloth : alk. paper)
1. Social behavior in animals—Mathematical models.   I. Boyd, Robert.   II. Title.
    QL775.M375 2007
    591.5601'5118—dc22

                                                    2006023442

⊗ The paper used in this publication meets the minimum requirements of the American National Standard for Information Sciences—Permanence of Paper for Printed Library Materials, ANSI Z39.48-1992.

Contents

CONTENTS vii

Preface

Imagine a field in which nearly all important theory is written in Latin, but most researchers can barely read Latin and certainly cannot speak it. Everyone cites the Latin papers, and a few even struggle through them for implications. However, most rely upon a tiny cabal of fluent Latin-speakers to develop and vet important theory.

Things aren't quite so bad as this in evolutionary biology. Like all good caricature, it exaggerates but also contains a grain of truth. Most of the important theory developed in the last 50 years is at least partly in the form of mathematics, the "Latin" of the field, yet few animal behaviorists and behavioral ecologists understand formal evolutionary models. The problem is more acute among those who take an evolutionary approach to human behavior, as students in anthropology and psychology are customarily even less numerate than their biology colleagues.

We would like to see the average student and researcher in animal behavior, behavioral ecology, and evolutionary approaches to human behavior become more fluent in the "Latin" of our fields. Increased fluency will help empiricists better appreciate the power and limits of the theory, and more sophisticated consumers will encourage theorists to address issues of empirical importance. Both of us teach courses to this end, and this book arose from our experiences teaching anxious, eager, and hard-working students the basic tools and results of theoretical evolutionary ecology.

How to use this book

While existing theory texts are generally excellent, they often assume too much mathematical background. Typical students in animal behavior, behavioral ecology, anthropology, or psychology had one semester of calculus long ago. It was hard and didn't seem to have much to do with their interests, and, as a result, they have forgotten most of it. Their algebra skills have atrophied from disuse, and even factoring a polynomial is only an ancient memory, as if from a past life. They have never had proper training in probability or dynamical systems. In our experience, these students need more hand-holding to get them started.

This book does more hand-holding than most, but ultimately learning mathematical theory is just as hard as learning a foreign language. To this end, we have some suggestions. In order to get full use of this book, the reader should:

1. Read the book in order. Each chapter builds on the last. After the first three chapters, it will be easier to jump ahead to topics of individual interest.

2. Work through examples within each chapter. Math is like karate: it must be learned by doing. If you really want to understand this stuff, you should work through every derivation and understand every step. Particularly involved derivations have been set in boxes, however, so that you can read the chapters at two levels, for general understanding and for skill development.

3. Work all the problems. At the end of each chapter, we provide a set of problems that elaborate upon the material in the chapter and allow the reader to practice the techniques we describe. The solutions to these problems are found at the end of the book. We encourage readers to attempt all the problems before looking at the solutions. We know these problems are solvable with only the tools presented in this book because our

own students have solved them. Some of the problems
are subtle, however, and we encourage you to collab-
orate with your friends and fellow students when you
attempt them.

4. Be patient, both with the material and yourself. Learning to understand and build formal evolutionary models is much like learning a language. This book is a
first course, and it can help the reader understand the
grammar and essential vocabulary. It is even a pass-
able pocket dictionary. However, only practice and use
of the language will build fluency. Students sometimes
jump to the conclusion that they are stupid because
they do not immediately understand a theoretical paper of interest. This is not justified. No one quickly
learns to comprehend and produce mathematical arguments, any more than one can quickly become fluent in
Latin.

The foundation of the foundations

Some reviewers of this book complained that some of the
material is old fashioned. True, nearly all of Chapter 2 is
30 years old. The derivation of Hamilton's rule in Chapter 3
is hardly new either, at least to professional theorists. Most
of Chapter 4 is old hat to mathematical social scientists, although biologists are unlikely to know much of it.

We think the vintage of the models is irrelevant for two
reasons. One reason is that these models still have currency in
important debates. But even if this material were no longer
relevant to debates within the field (eventually this should
be true), it would still be worth learning. Modern physi-
cists do the bulk of their day-to-day work in mechanics using
Lagrangian dynamics, a system more general and powerful
than the classical Newtonian dynamics most of us learned
in secondary school or college. Nonetheless, all physics programs teach Newtonian methods first because they are useful and easy to understand. In the same way, the simplest

game-theoretic and evolutionary models are worth knowing, even if, some day, they will no longer be applied in the day-to-day work of evolutionary theorists.

While this book provides a foundation for understanding the construction and analysis of formal models, it does not treat any of the topics exhaustively. At the end of each chapter, we provide a brief Guide to the Literature that directs interested readers to deeper treatments of the material.

Why so much algebra?

This book has lots of algebra. Much more, in fact, than many more-advanced texts in evolutionary theory. The reason is that beginning students need it. If you really want to learn this material, you have to follow the derivations one mathematical step at a time. Students that haven't done much math in a while often have a shaky grasp on how to do the algebra; they make lots of mistakes, and, even more important, don't have much intuition about how to get from point A to point B in a mathematical argument. All this means that they can find themselves stuck, unable to derive the next result, and unsure of whether they don't understand something fundamental or the obstacle is just a mathematical trick or algebraic error. Many years of teaching this material convinces us that the best remedy is to show lots of intermediate steps in derivations.

Acknowledgments

A number of people gave us feedback and advice during the long evolutionary process of writing this book. We have not always followed their advice, but we always appreciated it. These include Sam Bowles, Ford Denison, Thomas Dudey, Herb Gintis, Rick Grosberg, Mark Grote, Joe Henrich, Rufus Johnstone, Hanna Kokko, Mike Loeb, John McNamara, Karthik Panchanathan, Aimée Plourde, Andrew Pomiankowski, Peter Rodman, Joan Silk, Pontus Strimling, Masanori

Takezawa, Peter Todd, Eva van den Broek, Annika Wallin, and Franz Weissing.

We are particularly grateful to all of our students who endured various primitive forms of the book, at UCLA and UC Davis. Their reactions, both positive and negative, formed the most valuable feedback possible.

Chapter 1

The Theoretician's Laboratory

Mathematical models and the tools used to analyze them constitute the theoretician's laboratory. Simple mathematical models are experiments aimed at understanding the causal relationships that drive important natural phenomena. Theoreticians in evolutionary biology use a variety of tools to study such models, divining their secrets to reveal how interactions that take place over long time spans shape the evolution of behavior. These models are almost always too simple to make accurate predictions or even accurately represent how any real behavior evolves. Nonetheless, they have proven to be extremely valuable because they help us understand processes too complex to grasp by verbal reasoning alone.

Consequently, evolutionary biology has been greatly influenced by the use of such "experiments." In order to read the primary literature in evolutionary biology or animal behavior, you have to be able to really understand mathematical models. Unfortunately, few programs adequately prepare even advanced graduate students to read and understand mathematical theory. Even when research is largely empirical, predictions are often derived from mathematical models, and students of animal behavior—be the animals humans or otherwise—are handicapped by not fully understanding the theory from which the empirical work derives. What makes it worse is that having a good education in mathematics or probability, while certainly helpful, isn't enough to understand evolutionary models. One needs to acquire a unique toolbox of ideas for dealing with formal population dynamics.

As economists, anthropologists, and others outside biology have been drawn to this body of theory over the last few decades, the problem has become more acute. Now social scientists of many flavors are struggling with mathematical theories they never encountered in their own fields. Often, we are told that someone has shown some result mathematically, and we will just have to accept it. Sometimes, this leads to uncritical adoption of theory. Other times, it leads to uncritical rejection. Neither is an acceptable state of affairs. Both kin selection (Chapter 3) and reciprocal altruism (Chapter 4), topics central to the evolution of social behavior, are cases where the gulf between theorists and consumers of theory is far too wide.

This book aims to help make the student or professional researcher in biology or the social sciences conversant in the language of the evolutionary theory of social behavior. It assumes that the student can do algebra and has had an introduction to calculus but maybe has forgotten much of it. It covers many of the basic topics: contests and conflicts, kin selection, reciprocity, signaling, multilevel selection, sex allocation, and sexual selection. At the same time, readers will be introduced to many of the typical mathematical tools that are used to analyze evolutionary models. At the end of each chapter, we provide a set of problems that elaborate upon the material in the chapter and allow the reader to practice the techniques we describe. The solutions to these problems are found at the end of the book. Those who work through the book and do all the problems should have a reasonable grasp on the fundamental concepts that underly modern biology's understanding of the evolution of behavior, be able to intelligently and critically read papers in journals like *Animal Behaviour* or *Proceedings of the Royal Society, Series B*, and with a little more practice maybe start building models of their own.

1.1 The structure of evolutionary theory

When Darwin left for his voyage around the world on the *Beagle*, he took with him the first volume of Charles Lyell's *Principles of Geology*. Somewhere in South America, he received the second volume by post. Lyell is famous for, among other things, never accepting Darwin's account of evolution by natural selection, apparently because of his religious beliefs. There is a sad irony here because Lyell's work played a crucial role in the development of Darwin's thinking. In fact, in some ways Lyell's principle of uniformitarianism is as central to Darwinism as is natural selection.

Before Lyell, it was common to explain the features of the earth's geology in terms of past catastrophes: floods, earthquakes, and other cataclysms. In contrast, Lyell tried to explain what he observed in terms of the cumulative action of processes that he could observe every day in the world around us—the sinking of lands, the building up of sediments, and so on. By appreciating the accumulated small effects of such processes over long time spans, great changes could be explained.

Darwin took this idea and applied it to populations of organisms. Darwin was a good naturalist and knew a lot about the everyday lives of plants and animals. They mate, they give birth, they move from one place to another, and they die. Darwin's great insight was to see that organisms vary and that the processes of their lives affect which types spread and which vanish. The key to explaining long-run change in nature was to apply Lyell's principle of uniformitarianism to populations. By keeping track of how the small events of everyday life change the composition of populations, we can explain great events over long time scales.

Biologists have been thinking this way ever since Darwin, but it is still news in some fields. Are people products of their societies or are societies products of people? The answer must be "both," but theory in the social sciences has tended to take one side or the other. In evolutionary models, this classical conflict between explanations at the level of the society (think Durkheimian social facts) and explanations at the level of individuals (think micro-economics) simply disappears. Population models allow explanation and real causation at both levels (and more than two levels) to exist seamlessly and meaningfully in one theory. We don't have to choose between atomistic and group-level explanations. Instead, one can build models about how groups of individuals can create population-level effects which then change individuals in powerful ways. This aspect of evolutionary theory gives it great power to explain the evolution of behavior in both people and other animals.

1.2   The utility of simple models

The models you will analyze in this chapter, and in fact every model we consider in this book, are much simpler than the real systems that they represent. The populations are infinite (or at least very large), the organisms that populate them have highly stylized trait inheritance and reproduction, the generations are discrete and nonoverlapping, stochastic effects are largely nonexistent, and gene interactions are wholly absent.

Simple evolutionary models never come close to capturing all the detail in any real species or situation. Instead, game theory trades specificity and detail for tractability and clarity. Models are like maps—they are most useful when they contain the details of interest and ignore others. A map of subways helps a person to navigate by subway. A map of streets helps on the surface. A map with both subways and streets would be too crowded and confusing for either task. A passage from Lewis Carroll's *Sylvie and Bruno Concluded*[1] illustrates this point:

> "What do you consider the largest map that would be really useful?"
>
> "About six inches to the mile."
>
> "Only six inches!" exclaimed Mein Herr. "We very soon got six yards to the mile. Then we tried a hundred yards to the mile. And then came the grandest idea of all! We actually made a map of the country, on the scale of a mile to the mile!"
>
> "Have you used it much?" I enquired.
>
> "It has never been spread out, yet," said Mein Herr: "The farmers objected: they said it would cover the whole country, and shut out the sunlight! So now we use the country itself, as its own map, and I assure you it does nearly as well."

Game-theoretic models are simple maps for understanding the consequences of a small number of key assumptions. Despite their deliberate myopia, and perhaps as a result of it, they have proven quite useful in a number of fields, especially in the study of social behavior. Their simplicity limits their power, but it also allows them to do things for us that the real world cannot. For example, we can fit them in our pockets.

The models we will explore in this book are only as complicated as they must be to capture the structure of their problems. They always omit factors which are known to matter in the real world. Such omissions are intentional. It is accepted practice to control for variables in psychology or economics experiments, variables which are known matter in many important situations. But ignoring some (most!) variables allows experimenters to examine the effects of other variables in isolation. Simple models are much the same.

They are not meant to be replicas of the real world, but rather to help us understand it, one piece at a time.

Simple models can aid our understanding of the world in several ways.

Existence proofs.   There is usually a very large number of possible accounts of any particular biological or social phenomenon.   And since much of the data needed to evaluate these accounts are lost or impossible to collect, it can be challenging to narrow down the field of possibilities.   But models which formalize these accounts tell us which are internally consistent and when conclusions follow from their premises. With purely verbal arguments about evolutionary processes, it is all too often the case that our conclusions do not follow from our assumptions.   Unaided reasoning about the mass effects of many weak forces operating over many generations has proved to be hazardous. Formalizing our arguments helps us understand which stories are possible explanations.[2] They provide proof that some candidate set of processes *could* explain the observations of interest.

Aid communication.   Formal models are much easier to present and explain.   In almost every field in which theory has been important—philosophy, the natural sciences, economics—formal systems for communicating and inspecting theory have arisen. This is partly for the reasons just explained, but formal models are also important because they give us the ability to clearly communicate what we mean. The looseness of verbal models sometimes allows scholars to argue for years about their implications.   Such wrangling is much rarer in mathematical disciplines because you have to precisely specify how every piece of the theory relates to every other piece. There is comfort in vagueness, and formal theory allows for little comfort. Sometimes, being forced to spell out an idea in detail is enough to make the formal exercise worth our time.

Counterintuitive results. Simple models often lead to surprising results. If such models always told us what we thought they would, there would be little point in constructing them. Instead, as any practicing theorist can tell you, models often surprise us, producing effects and interactions we hadn't imagined prior to constructing the model. These unanticipated effects lead to additional theory construction and data collection. They take our work in directions we would have missed had we stuck to verbal reasoning, and they help us to understand features of the system of study that were previously mysterious.

Prediction. Simple formal models can be used to make predictions about natural phenomena. In most cases, these predictions are qualitative. We might expect more or less of some effect as we increase the real-world analogue of a particular parameter of the model. For example, a model of voter turnout might predict that turnout increases in times of national crisis. The precise amount of increase per amount of crisis is very hard to predict, as deciding how to measure "crisis" on a quantitative scale seems like a life's work in itself. Even so, the model may provide a useful prediction of the direction of change. These sorts of predictions are sometimes called *comparative statics*. And even very simple models can tell us whether some effect should be linear, exponential, or some other functional form. All too often, scientists construct theories which imply highly nonlinear effects, yet they analyze their data with linear regression and analysis of variance. Allowing our models of the actual phenomenon to make predictions of our data can yield much more analytic power.[3] In a few cases, such as sex allocation, models can make very precise quantitative predictions that can be applied to data.

But keep in mind that even when direct prediction from a model seems unlikely, models can direct our attention to things in the real world we should be measuring but had previously ignored. New data will then lead us to revise our models, leading in turn to more data collection. The relationship between models and data is not a one-way street. Data inspire and challenge theories just as often as theories inspire data collection. Or at least they should.

1.3   Why not just simulate?

There is a growing number of modelers who know very little about analytic methods. Instead, these researchers focus on computer simulations of complex systems. When computers were slow and memory was tight, simulation was not a realistic option. Without analytic methods, it would have taken years to simulate even moderately complex systems. With the rocketing ascent of computer speed and plummeting price of hardware, it has become increasingly easy to simulate very complex systems. This makes it tempting to give up on analytic methods, since most people find them difficult to learn and to understand.

There are several reasons why simulations are poor substitutes for analytic models.

Equations talk. Equations—given the proper training—really do speak to you. They provide intuitions about the reasons an evolutionary system behaves as it does, and these reasons can be read from the expressions that define the dynamics and resting states of the system. Analytic models therefore *tell us* things that we must infer, often with great difficulty, from simulation results. Analytic models can provide proofs, while simulations provide only a collection of examples.

Sensitivity analysis. It is difficult to explore the sensitivity of simulations to changes in parameter values. Parameters are quantities that specify assumptions of the model for a given run of the simulation—things like population size, mutation rate, and the value of a resource. In analytic models, the effects of these changes can be read directly from equations or by using various analytic techniques. In simulations, there are no analogous expressions. Instead, the analyst has to run a large number of simulations, varying the parameters in all combinations. For a small number of parameters, this may not be so bad. But let's assume a model has four parameters of interest, each of which has only 10 interesting values. Then we require 10^4 simulations. If there are any stochastic effects in the model, we will need maybe 100 or 1000 or 10,000 times as many.

To see how time-consuming this can be, suppose we have a fairly fast computer that can complete a thorough single run of a simulation in 1 second. This may seem like a long time to run a simulation, but a nondeterministic, agent-based simulation can take a long time to run out enough generations to get a good picture of any steady state or dynamic cycles. Now, we will need $10^4 = 10,000$ different combinations of parameter values to entirely map out the parameter space, at the given intervals. Suppose we require 1000 simulations at each combination, to reliably estimate the mean outcome in each case. Now we are looking at 10^7 simulations. At 1 second each, that's almost 116 days of computer time. No doubt computers will one day be fast enough to do 100 or more such simulations per second, but until then computer time is a major constraint.

In practice, people often ignore large portions of the parameter space, and so, many fewer simulations may be needed. There are often obvious and good reasons to do so. But even when large portions of the parameter space can be safely neglected, managing and interpreting the large amounts of data generated from the rest of the combinations can be a giant project, and this data-management problem will remain no matter how fast computers become in the future. Technology cannot save us here. When simple analytic methods can produce the same results, simulation should be avoided, both for economy and sanity.

Computer programs are hard to communicate and verify. There is as yet no standard way to communicate the structure of a simulation, especially a complicated "agent-based" simulation, in which each organism or other entity is kept track of independently.[4] Often, key aspects of the model are never mentioned at all. Subtle and important details of how organisms reproduce or interact can remain very murky indeed. In contrast, analytic models have benefitted from generations of notational standardization, and even unmentioned assumptions can be read from the expressions in the text. Thus it is much easier for other researchers to verify and reproduce modeling results in the analytic case. Bugs are all too common, and simulations are rarely replicated, so this is not a minor virtue.

Overspecification. The apparent ease of simulation often tempts the modeler to put in every variable which might matter, leading to complicated and uninterpretable models of an already complicated world. Surprising results can emerge from simulations, effects that we cannot explain. In these cases, it's hard to tell what exactly the models have taught us. We had a world we didn't understand and now we have added a model we don't understand.

If the temptation to overspecify is resisted, however, simulation and analytic methods complement each other. Each is probably most useful when practiced alongside the other. There are plenty of important problems for which it is simply impossible to derive analytic results. In these cases, simulation is the only solution. And many important analytic models can be specified entirely as mathematical expressions but cannot be solved, except numerically (we present an example in Chapter 8). For these reasons, we would prefer formal and simulation models be learned side by side. However, most people find learning formal modeling hard enough without learning computer programming on top of it. So while we will sometimes comment on the complementary value of simulation, for the most part it's pencil to paper, chalk to board, and full steam ahead from here on out.

1.4 A model of viability selection

Enough abstraction, let's get to work. In this section, we will build a simple model of evolutionary changes that result from variation in probability of survival—a form of natural selection that populations geneticists call *viability selection*. Along the way, we'll see how to go from a specification of the life cycle of an organism to deducing the long-term effects on the genes in the population. At the same time, we'll derive some useful expressions that we will use again and again and also learn some useful ways to visualize and interpret evolutionary models.

In any evolutionary model, the minimum we need is:

1. A description of the *population*: This describes how many individuals there are and how the social organization is structured.

2. A set of *heritable variants*: In genetic evolution, this usually includes different genotypes; in cultural evolution we speak of cultural variants, information stored in brains and transmitted from one generation to the next.

3. A *life cycle*: A description of the events in individual lives that affect the survival and proliferation of heritable variants.

The first step in building a model is to derive a mathematical representation of how the events in the life cycle change the distribution of heritable variants in the population over one time period (usually, but not always, one biological generation). Then we want to figure out what happens over the long run, as the events of the life cycle take place over many time periods. As we will see, there are a variety of techniques for accomplishing the second step.

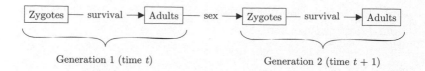

Generation 1 (time t)   Generation 2 (time $t + 1$)

Figure 1.1: Life cycle assumed in the viability selection model. At a given time t, zygotes in the population mature into adults, if they survive. Adults then mate and produce zygotes at time $t + 1$ who undergo the same process.

1.4.1 Selection for survival

Assume there is a population of n individuals, where n is large enough that we can ignore sampling variation. These individuals are haploid, meaning that there is no sex or recombination. There are two genotypes, A and B. Generations are discrete, meaning that individuals go through their life cycle in lock step—everyone is born at the same time, and everyone reproduces at the same time. Suppose n zygotes exist at some time t. Some zygotes manage to survive the slings and arrows of natural selection and mature into adults—A types are more likely to survive than B types. Surviving adults then reproduce, creating a population of zygotes at a time $t + 1$. Its often useful to represent the life cycle using a diagram like that shown in Figure 1.1.

As with each chapter in this book, we summarize definitions of the symbols used in this chapter in a box at the end of the chapter. Formal theory often requires more symbols than can be easily remembered. You can refer to this box as you work through the model in order to make sure you understand each step.

In most of the models we explore in this book, we keep track of the proportions or frequencies of different genotypes or strategies in a population. These frequencies are numbers between zero and one that simply tell us what fraction of the population is of a certain type. They are examples of what mathematicians who study dynamical systems call *state variables*, quantities that describe the state of the system and are sufficient to predict the state at future times. In population-genetic models the number of state variables is often one less than the number of alternative alleles or strategies in the population. For example, if a population has only two types, A and B, then the frequency of A, p_A, tells us the portion of the population of type A. The proportion that is type B could be expressed as p_B or simply $(1 - p_A)$, since there are only two alleles in this example. In a system with three alternative types, we would need two state variables to specify all the frequencies: p_1, p_2, and $1 - p_1 - p_2$.

In this simplest of evolutionary models, only one evolutionary force is at work. Individuals with some genotypes are more likely to survive than others. In order to deduce the long-term effects of these events, we need to specify how individual genotypes survive from one time period to the next. In essence, we need to perform a census of the genotypes at time t and at time $t + 1$. Recall that there are two genotypes, A and B, in the population. Let p be the frequency of the A genotype in the population at time t. This is simply

$$p = \frac{\text{number of A zygotes}}{n},$$

where n is the number of individuals currently in the population. The above expression means that the number of A zygotes is equal to np. Consequently, the number of B zygotes is $n(1 - p)$. This means we are now keeping track of the differential evolution of types A and B in this population with a single number, p. Thus from stage to stage in the life cycle, the value of p may change, depending upon how events affect the different alleles. The variable p is our only state variable. By convention, p' (pronounced "p prime") represents the value of our state variable during the next stage of the life cycle. In calculus, the $'$ symbol sometimes means the derivative of a function. In this book, it will mean the value of the variable in the next time period or stage in the life cycle.

So far, we have only counted up the numbers of different genotypes among the zygotes, to get our state variable. No biology so far. Next, these zygotes will mature, and some will survive to adulthood, and this is where the biology comes in. Let $V(A)$ and $V(B)$ be the probabilities that a zygote of each type survives to adulthood (we will use V's throughout the book to refer to payoffs to types). For example, when $V(A) = 0.5$, half of all zygotes with the A genotype survive to be adults. With this assumption, we can do a census for the population of adults. The number of adults with genotype A will be

$$\text{number of } A \text{ adults} = npV(A).$$

The number of B adults is then $n(1-p)V(B)$. The frequency of genotype A among adults, p', is given by

$$p' = \frac{\text{number of } A \text{ adults}}{\text{number of } A \text{ adults} + \text{number of } B \text{ adults}}$$

$$= \frac{npV(A)}{npV(A) + n(1-p)V(B)}.$$

After selection, the remaining adults mate and reproduce. In this model, let's keep reproduction simple. Each individual reproduces asexually, producing z zygotes, regardless of genotype. When this is true, p'', the frequency of genotype A among zygotes at time $t + 1$, is given by

$$p'' = \frac{(\text{number of } A \text{ adults})z}{(\text{number of } A \text{ adults})z + (\text{number of } B \text{ adults})z} = p'.$$

Since all genotypes produce the same average number of offspring, reproduction in itself does not change the frequencies of A and B genotypes.

Given the work above, we can write an expression for the frequency of genotype A after one generation:

$$p'' = \frac{pV(A)}{pV(A) + (1-p)V(B)}. \tag{1.1}$$

Box 1.1 Derivation of viability selection difference equation

Begin by subtracting the frequency of type A at the start of the generation, p, from each side of the recursion:

$$\Delta p = p'' - p = \frac{pV(A)}{pV(A) + (1-p)V(B)} - p.$$

Multiply the top and bottom of the far-right term by $pV(A) + (1-p)V(B)$:

$$\Delta p = \frac{pV(A)}{pV(A) + (1-p)V(B)} - \frac{p(pV(A) + (1-p)V(B))}{pV(A) + (1-p)V(B)}.$$

After simplifying, we obtain

$$\Delta p = \frac{p(1-p)\big(V(A) - V(B)\big)}{pV(A) + (1-p)V(B)}.$$

We commonly write the denominator $pV(A) + (1-p)V(B)$ and expressions like it as $\bar{w}$, which is the average fitness in the population. This gives us a final difference equation in the main text.

We normally call an expression of this kind a *recursion*, and it allows us to apply the per-generation effects of evolutionary forces over any number of generations. To see why, consider a case where we know the frequency of genotype A in generation 1 (let's call this p_1). Then we can calculate the frequency of genotype A in generation 2 (p_2) as

$$p_2 = \frac{p_1 V(A)}{p_1 V(A) + (1 - p_1)V(B)}.$$

The frequency of genotype A in generation 3 (p_3) is likewise calculated by substituting p_2 into the recursion.

Often, it is useful to represent these events using a *difference equation*. Instead of specifying the new frequency after one generation, a difference equation yields the *change* in the frequency after one generation. In Box 1.1, we demonstrate how to convert the recursion above into the difference equation

$$\Delta p = p(1 - p) \frac{V(A) - V(B)}{\bar{w}}.$$

This recursion and difference equation are commonly called *replicator dynamics* for viability selection. We will see them again and again, so understanding what they say is important. This difference equation reveals something interesting about how natural selection (currently the only force in our model) changes genotype frequencies. The first part, $p(1-p)$, is the variance in genotypes in the population (you will prove this later in the book). When either genotype is common, this product is small. The variance is maximized when both genotypes are equally common ($p = 0.5$). This means that natural selection is strongest when variance is maximized. Natural selection is a culling process, and it changes genotypes most quickly when there is more to cull. At the limit, when there is no variation, natural selection cannot change genotype frequencies at all. We will see variance terms like this crop up again and again.

The second part, $\frac{V(A)-V(B)}{\bar{w}}$, is simply the proportional increase or decrease of genotype A compared to genotype B. If genotype A is more fit than genotype B, this term will be positive, and the frequency of genotype A will increase each generation as long as there is still some variation in genotypes ($p(1 - p) > 0$). Likewise, when $V(A) < V(B)$, the frequency of genotype A will decrease each generation.

A final concern is what happens when fertility, the z's in the model above, differs among the types. We derive a general combined viability/fertility recursion in Box 1.2.

1.5   Determining long-term consequences

The recursion is a mathematical representation of how events in the lives of individuals change the genetic composition of the population over one generation. Evolutionists want to figure out what will happen in the long run if these processes

---

Box 1.2 Selection on fertility and viability

What happens when different types produce different numbers of zygotes? This would be selection on fertility, and is in fact what many of the models in this book assume. If we let $z(A)$ and $z(B)$ be the average number of zygotes produced by each type, the viability recursion (Equation 1.1) becomes

$$p'' = \frac{pV(A)z(A)}{pV(A)z(A) + (1-p)V(B)z(B)}.$$

This implies that the fitness of each type is the product of its survival and reproductive success. Thus a combined viability/fertility selection recursion can be written using $W(A) = V(A)z(A)$ and $W(B) = V(B)z(B)$:

$$p'' = \frac{pW(A)}{pW(A) + (1-p)W(B)} = p\frac{W(A)}{\bar{w}}.$$

The combined difference equation is naturally

$$\Delta p = p(1-p)\frac{W(A) - W(B)}{\bar{w}}.$$

---

are repeated generation after generation. There are a number of ways to do this. For very simple recursions like the one above, it is possible to calculate (or rather, guess at) an explicit solution for the number or frequency of genotype A at any time t. However, this will not be the case for most of the other models we examine. For even slightly more complicated recursions, explicit solutions are almost always impossible. Therefore, in this section we present several ways to go about deriving long-term consequences. First, we explore an explicit solution. In this case, the explicit solution is possible and even of some interest to us. Second, we demonstrate how to visualize the dynamics of these systems and derive outcomes graphically. Third, we explain how to derive long-term consequences in general by linearizing the recursion.

Box 1.3 Explicit solution to viability recursion
Notice that the number (not the frequency) of A alleles at time $t + 1$ is just

$n_{A,t+1} = (\text{\# zygotes per individual}) \times n_{A,t} \times (A \text{ type survival})$
$n_{A,t+1} = z n_{A,t} V(A).$

The number of B alleles is irrelevant in this computation, since the fitness of both types is a constant (it isn't affected by density in any way). So the number of A alleles at time $t + 2$ is just

$$n_{A,t+2} = z n_{A,t+1} V(A) = z\{z n_{A,t} V(A)\} V(A)$$
$$= n_{A,t}\big(z V(A)\big)^2.$$

and at time $t + 3$ it is

$$n_{A,t+3} = z n_{A,t+2} V(A) = z\{n_{A,t}(z V(A))^2\} V(A)$$
$$= n_{A,t}\big(z V(A)\big)^3.$$

You can probably see by now that at any time t the number of A alleles will be
$$n_{A,t} = n_{A,0}\big(z V(A)\big)^t,$$

where $n_{A,0}$ is the number of A alleles in the first time period. The number of B alleles at any time t is similarly

$$n_{B,t} = n_{B,0}\big(z V(B)\big)^t.$$

So the frequency of A alleles at any time t is just

$$p_t = \frac{n_{A,t}}{n_{A,t} + n_{B,t}} = \frac{n_{A,0} V(A)^t}{n_{A,0} V(A)^t + n_{B,0} V(B)^t}.$$

Finally, dividing the numerator and denominator by a factor $n_{A,0} V(A)^t$ yields the function in the main text (Equation 1.2).

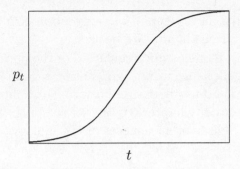

Figure 1.2: The long-term consequences of selection, as derived from Equation 1.2. The frequency of A alleles increases slowly at first, most quickly when the variance is maximized at $p = 0.5$, and again slows down before reaching 1.

1.5.1   Explicit solution

An explicit solution is an expression that tells us the frequency of each type at any time t. We explain in Box 1.3 how to find the explicit solution in this case. The function giving the frequency of A alleles at any time t is

$$p_t = \frac{1}{1 + \left(\dfrac{n_{B,0}}{n_{A,0}}\right)\left(\dfrac{V(B)}{V(A)}\right)^t}. \tag{1.2}$$

This is the logistic growth equation commonly used in ecology to model population growth.

Let's plot the frequency over time, given some values for $V(A)$ and $V(B)$. Plotting the frequency over time is often useful for understanding the ongoing evolutionary process. Let the initial frequency of genotype A be 1% ($p_0 = 0.01$), $V(A) = 0.8$, and $V(B) = 0.5$. Figure 1.2 shows the value of p_t for each generation, using the explicit solution in Equation 1.2. Natural selection acts very slowly at first, since there is not much variation in the population. As p increases, selection acts more strongly. Finally, selection slows down again as genotype Bs are largely removed from the population.

Eventually, in a finite population, all genotype Bs will disappear, and $p = 1$ on the infinite horizon.

We haven't modeled any regulation of the population size here, and in general we won't. Population regulation is interesting, and sometimes it matters a lot exactly how it works (we take this up in Chapter 3), but we're going to focus our attention on mainly other aspects of the biology.

1.5.2   Equilibrium analysis

Explicit solution is hardly ever possible. So, for most models we have to do with a second best option, *equilibrium analysis*. This is the process of deriving values of the state variables (in this case there is only one state variable, p) for which the system does not change from one time period to another and determining if the system moves back to any of these values when perturbed a bit. *Equilibria* are states at which the system gets stuck, so once it gets to an equilibrium state, it stays there. Equilibria are *stable* if the system returns to them once perturbed a small amount. These perturbations can arise from mutation, migration, or other forces. We have to derive the equilibria and determine their stability because not all equilibria are stable. Both kinds of equilibria are of great interest. Stable equilibria are candidates for long-term evolutionary outcomes. Unstable equilibria are interesting, as they usually define the boundaries between *basins of attraction* of two or more stable equilibria. A basin of attraction of a stable equilbrium is the set of initial conditions which converge to that equilibria. When there is more than one equilibrium, we need to know which is the most likely evolutionary outcome, and the size of the basin of attraction is one (but not the only) measure of this. (It can also be that no equilibrium is stable, and the system changes forever. Such behavior is important in ecology but less so in the models in this book.) We'll have examples of all these things before long, so hang on while we work through it.

Equilibria in our model here can be found by deriving the frequencies at which $p'' = p$. We will label these equilibrium frequencies $\hat{p}$. It is often easiest to find these equilibria by setting the difference equation (Δp, remember) equal to zero, which implies that we are finding the value of p which leaves the frequency of p the same from one time period to the next. The equilibrium in our system lies where

$$\Delta p = \hat{p}(1 - \hat{p})\frac{V(A) - V(B)}{\bar{w}} = 0.$$

If any of the terms in this equation equals zero, the whole thing is zero, and we are at an equilibrium. The first two equilibria are easy to spot. When $p = 1$ or $p = 0$, the entire expression is zero. When either genotype exists alone, natural selection cannot change the genotype frequencies. The only other way for Δp to equal zero is when $V(A) = V(B)$. When both genotypes have the same probability of surviving to adulthood, natural selection does change the genotype frequencies. This situation isn't very interesting here, however, as it simply tells us that natural selection does not act in a population when all genotypes have the same fitness. In the following chapters (and one of your homework problems) we will study cases in which fitness depends on the composition population, and in such cases, this equilibrium will be very interesting to us. But not yet.

Taking the two meaningful equilibria, $\hat{p} = 0$ and $\hat{p} = 1$, let's ask when they are stable. Figure 1.3 plots the recursion from the model we derived above for all values of p from zero to one, for $V(A) > V(B)$. The thick line plots p' as a function of p as given by the recursion, and the dashed line is simply the graph of the set of points for which $p' = p$. Wherever the thick line and the dashed line intersect, there is an equilibrium. In this case, there are equilibria at zero and one.

Now let's investigate what happens when the system is at $\hat{p} = 0$ and we perturb it a bit away, so that p is slightly greater than zero (Figure 1.3(a)). Will the system go back to $\hat{p} = 0$ or will it run away? There is a simple graphical way

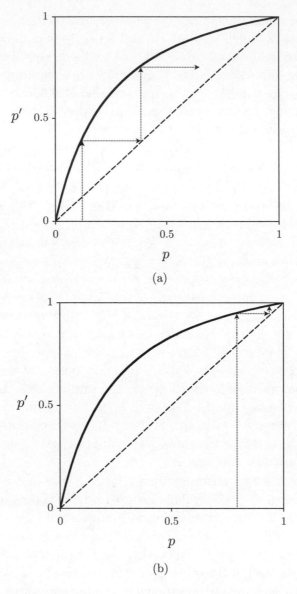

(a)

(b)

Figure 1.3: Graphical dynamics when $V(A) > V(B)$. (a) *An unstable equilibrium.* The system runs away toward the other equilibrium when perturbed a bit away from the equilibrium at $\hat{p} = 0$. (b) *A stable equilibrium.* In this case, the population returns to the equilibrium when perturbed a bit away from the equilibrium at $\hat{p} = 1$.

of answering this question, using graphs like the the one in Figure 1.3(a). Let's pick a value of p along the x-axis that is slightly greater than zero and draw a line upwards until it meets the curve for p'. We just traced one iteration of the recursion. To find where p goes from this point, we extend our line to the diagonal line where $p' = p$, then change direction again and meet the curve for p', as shown in Figure 1.3. We could continue this process all the way to $\hat{p} = 1$, which is where the system is heading. Thus, the equilibrium at $\hat{p} = 0$ is unstable. What about the equilibrium at $\hat{p} = 1$? Using the same technique (Figure 1.3(b)), we see that the system returns to $\hat{p} = 1$, so this equilibrium is stable.

This technique works for any continuous recursion. We'll encounter several more complex recursions very soon, and it is left as an exercise for the reader to verify that the graphical method here and the analytical method we're about to encounter lead to the same conclusions.

1.5.3 Linear stability analysis

The formal method for determining the stability of an equilibrium is to compute the derivative of the recursion and evaluate it at the equilibrium. If the resulting value is less than 1 and greater than -1, the equilibrium is stable. Why is this? Computing the derivative of a function tells us its rate of change at each point. In this context, the rate of change is the change in p' as a function of p. It is the slope of the line tangent to the recursion for a given value of p.

Consider now an imaginary recursion that is just the straight line $p' = p$. At every point on this function, the slope is 1. Every point is an equilibrium, and so there is no deterministic change at any point p. Thus you can see that when the slope is 1, there is no deterministic result after a small perturbation. The system will stick wherever it ends up after the perturbation.

Suppose at some point the slope is less than 1, however. This means the recursion passes *under* the line $p' = p$ on the right side of the equilibrium and *over* the line on the left side of it (Figure 1.4(a)). Now positive perturbations lead to diminishing values, until the system returns to the equilibrium. The same argument explains why negative perturbations also return to the equilibrium. Since the slope of this recursion is less than 1, the equilibrium is stable. Technically, the slope must also be greater than -1 in order for the equilibrium to be stable. Can you figure out why, using the graphical method?

For nonlinear recursions, we use the same tactic. If perturbations away from the equilibrium are small enough, then a linear estimation of the function at the equilibrium point is all we need to know in order to tell whether the equilibrium is stable (Figure 1.4(b)). In essence, for a curved recursion, we want to know if it looks like the recursion in Figure 1.4(a) in the region of equilibrium under question. If the slope in this region is less than 1 and greater than -1, then it's stable, at least for small perturbations in the region of the equilibrium.

So how do we do this mathematically? Here's the recipe:

1. Take the derivative of the general recursion (Equation 1.1 in this case) with respect to p. This gives us the slope of the recursion at any point p.

2. Find the value of the derivative when $p = \hat{p}$. This gives us the slope of the line tangent to the recursion at the equilibrium point $\hat{p}$.

3. If this value is less than 1 and greater than -1, the equilibrium is stable.

Step 1.   Let's take the recursion we derived previously and find its derivative. We want

$$\frac{dp'}{dp} = \frac{d}{dp}\left(\frac{pV(A)}{\bar{w}}\right) = \frac{V(A)V(B)}{\bar{w}^2}.$$

In Box 1.4, we remind the reader how to find this result.

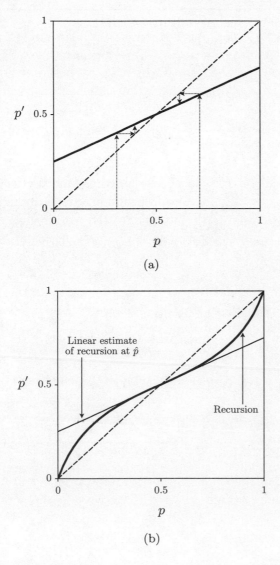

(a)

(b)

Figure 1.4: Linear estimation of evolutionary dynamics. (a) *A linear recursion.* The slope of the recursion at the equilibrium is less than 1, so it is stable, as shown by the graphical dynamics. (b) *A nonlinear recursion.* The slope of the recursion at the equilibrium (shown by the line tangent to the curve at $p = 0.5$) is less than 1, so this equilibrium is also stable.

Box 1.4 Derivative of viability recursion

To calculate the derivative $\frac{d}{dp}\frac{pV(A)}{\bar{W}}$, recall the quotient rule:

$$f(x) = \frac{g(x)}{h(x)} : \frac{d}{dx}f(x) = \frac{\frac{d}{dx}g(x) \times h(x) - g(x) \times \frac{d}{dx}h(x)}{h(x)^2}.$$

Thus in our case, $f(p)$ is the entire recursion and

$$g(p) = pV(A),$$
$$h(p) = \bar{W} = pV(A) + (1 - p)V(B).$$

The derivative of $pV(A)$ is just $V(A)$. This is true only because $V(A)$ does not depend upon p. If it did, the derivative would be more complicated. The derivative of $h(p)$ is similarly

$$\frac{d}{dp}h(p) = V(A) - V(B).$$

Combining the above expressions gives

$$\frac{d}{dp}\left(\frac{pV(A)}{\bar{W}}\right) = \frac{\{V(A)\}\bar{W} - pV(A)\{V(A) - V(B)\}}{\bar{w}^2}$$

$$= V(A)\frac{\{pV(A) + (1 - p)V(B)\} - p\{V(A) - V(B)\}}{\bar{w}^2}$$

$$= \frac{V(A)V(B)}{\bar{w}^2}.$$

Step 2.   Let's consider the first equilibrium, at $\hat{p} = 0$. We find the value of the derivative above when $p = \hat{p} = 0$:

$$\left.\frac{dp'}{dp}\right|_{p=0} = \frac{V(A)V(B)}{V(B)^2} = \frac{V(A)}{V(B)}.$$

Step 3.   Then we ask when the slope is less than 1 and greater than -1. Since neither $V(A)$ nor $V(B)$ can be negative, the slope is always greater than -1. What about less than 1?

$$\frac{V(A)}{V(B)} < 1$$

when

$$V(A) < V(B).$$

When $V(A) > V(B)$, the slope will be greater than one, and so the equilibrium at $\hat{p} = 0$ is unstable. When the opposite is true, $V(A) < V(B)$, the quotient is less than one, and so it is stable. Happily, this makes sense: when the fitness of the A allele is greater, it will increase in frequency after it is introduced in small numbers. If instead B is more fit, selection will remove any rare A alleles after introduction. A simple calculation shows a symmetrical result for the other equilibrium, at $\hat{p} = 1$.

1.6 Nongenetic replication

In some species, many important behavioral variants are transmitted by social learning rather than by genetic transmission. We might wonder if the replicator dynamics we derived above are applicable to such cases. Here, we will introduce you to a very useful tool called a *mating table*, as well as derive the dynamics for one form of social learning. We will see that social learning dynamics can be very similar to viability selection dynamics for simple genetic models.

A mating table is used in population-genetic models to calculate the outcomes of different possible matings in a population. For any particular set of parents, the mating table gives the probabilities of each type of offspring. Using these probabilities, we can then write a set of recursions for the system. The nice thing about mating tables is that they are very general tools. We can use them for almost all problems which will concern us in this book. And like many tools, they break a complex problem into smaller manageable components.

Here, we will construct a "mating" table that calculates the probabilities of different behaviors being imitated, given that an individual encounters a certain set of models. Suppose individuals behave, receive payoffs, and then compare their own payoff to their estimate of the payoff of another, random individual. If the other individual has a higher pay-off, the individual imitates the other individual's behavior. The probability that an individual acquires another behavior is proportional to the difference between the two payoffs.

We begin a mating table by writing down the types of possible "encounters" we are interested in. Let's assume here that there are only two possible alternative behaviors, A and B. Then there are four possible combinations: A and A, A and B, B and A, and B and B. For each of these possible combinations of self and other, we now write the probability that this combination occurs. Let p be the frequency of behavior A. Since the other individual is chosen at random, the probability that the other individual practices behavior A is p and the probability that he practices behavior B is $1 - p$. The chance that our focal individual has behavior A is likewise p, and the chance of B is likewise $1 - p$. This gives us the familiar probabilities for each type of interaction, shown in Table 1.1.

When the number of a certain type in the population is expressed as a frequency, we can use this frequency to compute the probability of different types of encounters. Given that only types A and B exist in a population and that the frequency of type A is p, the chances of each type of two-person encounter are

$$\Pr(A, A) = p^2$$
$$\Pr(A, B) = 2p(1 - p)$$
$$\Pr(B, B) = (1 - p)^2.$$

This is exactly like random mating in The Hardy-Weinberg model, which many of you learned about during your first year in college. The conditional probabilities are even easier, assuming random interaction (which we often will). Given that an individual is type A, the chance she meets an individual of type A ($\Pr(A|A)$) is just p. Likewise, the chance she meets a type B ($\Pr(B|A)$) is just $(1 - p)$.

Now we write a column for each type the focal individual

can end up as. If this were a usual genetic model, we'd write down a column for each genotype of the offspring. In this case, we write a column for each behavioral alternative. In these columns, we write the probabilities that the interaction on that row produces the given behavioral alternative after social learning. In order write these probabilities, we need to be a bit more specific about what we are assuming.

We need an expression that produces in any A and B interaction, when the payoffs to behaviors A and B are equal, an equal chance of the individual acquiring either. If the payoffs are equal, the individual will be indifferent about which behavior he should imitate. We could achieve indifference by entering a $1/2$ in each cell for each A/B row (there are two A/B rows in the table). But then we need to bias this $1/2$ by the difference in payoff between behavior A and behavior B. Assume that an individual who is in a pair with both an individual with behavior A and one with behavior B acquires behavior B with probability

$$\Pr(B|A, B) = \frac{1}{2} + \beta\{V(B) - V(A)\}.$$

$V(A)$ and $V(B)$ are the payoffs to behaviors A and B, respectively. The parameter β scales the term in braces so that the total probability is always between zero and one. If we make beta small, this is easy enough. Essentially, β tells

Table 1.1: Mating table Step 1: Kinds of interactions and the probability of each.

Self	Other	Probability
A	A	p^2
A	B	$p(1-p)$
B	A	$(1-p)p$
B	B	$(1-p)^2$

Table 1.2: Mating table Step 2: Probabilities of acquiring each behavior. To write a recursion for the frequency of A, p, we multiply the probability in each row by $\Pr(A|\text{self}, \text{other})$ in each row. Then we add up each of these row products to get the total expected frequency of A's in the next time period.

Self	Other	Pr(self, other)	Pr(A|self, other)	Pr(B|self, other)
A	A	p^2	1	0
A	B	$p(1-p)$	$\frac{1}{2} + \beta\{V(A) - V(B)\}$	$\frac{1}{2} + \beta\{V(B) - V(A)\}$
B	A	$(1-p)p$	$\frac{1}{2} + \beta\{V(A) - V(B)\}$	$\frac{1}{2} + \beta\{V(B) - V(A)\}$
B	B	$(1-p)^2$	0	1

us how strongly imitation is biased by difference in payoff. Now look what happens when $V(B) > V(A)$. Then the total probability is greater than $1/2$, so odds are that the individual acquires the more profitable behavior. Likewise, $\Pr(A|A, B)$ is given by

$$\Pr(A|A, B) = \frac{1}{2} + \beta\{V(A) - V(B)\}.$$

Where did these expressions come from? We made them up, as descriptions of the outcomes of social learning. Part of the creative art in game theory is imagining such formal expressions for outcomes that capture our intentions and make the math simpler. The expressions above fulfill our desiderata of being random choice when there is no payoff difference between the individuals and increasing bias in imitation toward the higher payoff individual when they do differ. Let's fill out the mating table now, by entering these probabilities in the correct cells (Table 1.2). The probabilities in the last two columns in the table must sum to one for each row. A little algebra confirms that $\frac{1}{2} + \beta\{V(A) - V(B)\} + \frac{1}{2} + \beta\{V(B) - V(A)\} = 1$.

We've written the entire mating table, so now we can construct a recursion for the frequency of either behavior. We accomplish this by multiplying the probability of each type of interaction (from the third column) by the probability of the outcome type we are counting (either the fourth or fifth

---

Box 1.5 Deriving a difference equation from a mating table

We take each probability of a pairing in Table 1.2 and multiply that by the probability that the pairing produces an individual of type A:

$$p' = \Pr(A, A)\Pr(A|A, A) + \Pr(A, B)\Pr(A|A, B)$$
$$+ \Pr(B, A)\Pr(A|B, A) + \Pr(B, B)\Pr(A|B, B)$$

$$p' = p^2(1) + 2p(1 - p)\left(\frac{1}{2} + \beta\{V(A) - V(B)\}\right) + (1 - p)^2(0).$$

We've added together the two middle rows to produce the term that begins $2p(1 - p)$. To turn this into a difference equation, we just subtract p from both sides and simplify:

$$\Delta p = p' - p$$
$$= p^2 + 2p(1 - p)\left(\frac{1}{2} + \beta\{V(A) - V(B)\}\right) - p$$
$$= p(1 - p)2\beta\{V(A) - V(B)\}.$$

---

column), for all rows in the table. In Box 1.5, we show how to follow this procedure to compute the difference equation for behavior A:

$$\Delta p = p(1 - p)2\beta\{V(A) - V(B)\}.$$

This expression is often simply called *the replicator dynamic*, and it has been derived independently many times, in more or less general forms, in both economics and biology.[5] This is very similar to the viability selection dynamics we derived earlier. The change in behavior A depends upon the variance in behaviors in the population and the direction, and magnitude of change depends upon the payoffs to the different behaviors, just as in the simple haploid genetic model. Provided the population is very large, it turns out that many different assumptions about payoff-sensitive replication lead to similar replicator dynamics. This fact will be useful to us, since it will allow us to use the same game-theoretic tools, at least with some squinting, for models of both genetic and learned behaviors. Keep in mind, however, that it is always possible to derive unique replicator dynamics for a given model by using a mating table. We are not stuck with any particular replicator dynamic, although many different dynamics may lead to the same or very similar equilibria.

Guide to the Literature

Philosophy of model building. Wimsatt's 1987 essay on models is helpful for many, students and professionals alike. Practical advice on model formulation and application can be found in Hilborn and Mangel's 1997 book, *The Ecological Detective.* Both give reasons why the best model for a scientific purpose may not be the most realistic model. **Population genetics and ecology.** Roughgarden's 1979 text is now in print again (Roughgarden 1996) and remains a useful and clear introduction to formal evolutionary ecology, especially when it comes to the goals and procedures of analysis. Alan Hastings' 1997 primer on population ecology covers inter-specific and age-structured models that we essentially ignore. **Social learning dynamics.** The payoff-biased imitation recursion in this chapter is commonly employed in economics now. Many other assumptions about imitation are possible. The psychology of how people learn is surprisingly poorly understood. There are many successful models, but most tend to make very similar predictions. See Camerer 2003, Chapter 6, for a nice review and discussion. Highly nonlinear learning, like majority-rule conformity, produces quite different replicator dynamics than that we derived here. See Boyd and Richerson 1985 for models of other imitation rules.

Problems

1. Linear dynamics. Use the graphical method to sketch the trajectory for the recursion $x' = x + a(x-1)$ for $a = -1.5$, $a = -2.0$, and $a = -2.5$.

2. Two-way mutation. Natural selection isn't the only force which changes the frequencies of alleles, of course. Let's derive the dynamics for a simple model in which mutation slowly adds and subtracts alleles. Imagine a very large population with two alleles at a single locus, labeled 1 and 2. Let p represent the frequency of allele 1. Further more, let m_{12} be the proportion of allele 1's that become allele 2's each generation. Similarly, m_{21} is the proportion of allele 2's that become allele 1's each generation. These numbers are usually called mutation rates. Assume that mutation occurs during gamete production. The frequency of allele 1 after mutation is given by

$$p' = p - m_{12}p + m_{21}(1 - p).$$

A fraction m_{12} of allele 1's become allele 2's, and so are subtracted each generation. A fraction m_{21} of the allele 2's become allele 1's, and so are added each generation.

Show that this system has a single equilibrium at

$$\hat{p} = \frac{m_{21}}{m_{12} + m_{21}}.$$

Also prove that this equilibrium is stable.

3. Social learning and replicator dynamics. Imagine an organism capable of social learning that acquires a certain behavior by selecting two models at random and imitating with some probability the one with the more successful behavior. Suppose there are two alternative behaviors and that one is better on average, and this difference is observable, but with some error. Precisely, assume that individuals make mistakes in comparing payoffs at a rate e such that they acquire the more successful behavior $1 - e$ of the time and mistakenly acquire the less successful behavior e of the time. When both models are the same, assume individuals copy either model at random. Construct the mating table for this model and use it to show that the change in the frequency of the more profitable trait, p, is

$$\Delta p = p(1 - p)(1 - 2e).$$

4.   When fitness is not a constant. Sometimes, the payoff to a certain behavior or the fitness of a certain allele depends upon the frequency of one or more behaviors or alleles in the population. A well-known example of this is the human sickle-cell allele, which has highest fitness in the heterozygote genotype.

Let w_{AA}, w_{AS}, and w_{SS} be the fitnesses of each genotype. For a diploid system with two alleles A and S, the change in the frequency of the A allele after selection is given by

$$\Delta p = p(1 - p)\frac{w_A - w_S}{\bar{w}}.$$

(a) Write expressions for w_A, w_S, and $\bar{w}$.

(b) Show that this system has an equilibrium at

$$\hat{p} = \frac{w_{AS} - w_{SS}}{2w_{AS} - w_{AA} - w_{SS}}.$$

(c) Using Excel (or any other program you like), plot the dynamics of this system for all values of p, assuming first that $w_{AA} = 0.9$, $w_{AS} = 1$, and $w_{SS} = 0.5$ and then that $w_{AA} = 0.9$, $w_{AS} = 0.75$, and $w_{SS} = 0.5$. That is, produce two graphs.

(d) Use the graphical method to prove that the internal equilibrium is stable only when w_{AS} is greater than both w_{AA} and w_{SS}. Provide a verbal argument as to why this is true. How does the system behave when w_{AS} is smaller than both w_{AA} and w_{SS}?

5.   When environments vary with time. Consider a haploid population in which there are two genotypes A and B. This population lives in an seasonal environment with a wet season and a dry season. Individuals are born at the beginning of each season, reproduce during that season, and die before the next season begins. The A genotype has fitness 1 in both seasons, while B has fitness 0.2 in the dry season and fitness 2 in wet season. Show that the only stable equilibrium is a population in which all individuals are A. How can you reconcile this with the fact that B has higher average fitness?

6. Horizontal cultural transmission, again. Consider a population of individuals who live forever. They have one of two culturally transmitted behaviors, A and B. Each time period individuals accumulate payoffs $V(A)$ and $V(B)$, where $V(A) > V(B)$. Individuals are then paired at random and observe the behavior and payoff of the individual with which they are paired. If the other individual has a higher payoff, they switch with a probability proportional to β(own payoff — other's payoff). Let p be the frequency of A. Show that the change in p over on time period is $\Delta p = \beta(V(A) - V(B))$.

Notes

[1] 1893.

[2] The philosopher of biology Robert N. Brandon 1990 argues that such "how-possibly" stories play a crucial role in evolutionary biology.

[3] See Hilborn and Mangel 1997 for an accessible introduction to marrying simple models to data analysis.

[4] Turchin 2003.

[5] Gintis 2000 shows a derivation for any number of alternative behaviors.

Box 1.6 Symbols used in Chapter 1

A and B Alternative genotypes in the population

t Index of the time or generation

n Number of individuals in the population at the current time

p Frequency of A in the population

p' Frequency of A in the next stage of the life cycle

$\hat{p}$ Equilibrium value of p

Δp Change in the frequency of A from time t to time $t+1$

$V(A)$ Probability a type A individual survives to adulthood

$V(B)$ Probability a type B individual survives to adulthood

z Average number of zygotes produced by either type of individual

$\bar{W}$ Average fitness, defined as $pV(A) + (1-p)V(B)$

$z(A)$ Average number of zygotes produced by type A

$W(A)$ Fitness of type A, defined as $V(A)z(A)$

$\Pr(x, y)$ Probability of events x *and* y occurring together

$\Pr(x|y)$ Probability of event x, *given* event y

$\Pr(x|y, z)$ Probability of event x, *given* events y *and* z

β A constant that determines the strength of imitation

Chapter 2

Animal Conflict

Resources are often scarce, and scarcity often leads to conflict. The form of conflict ranges from subtle supplants in which one animal departs at the other's approach to escalated fights in which participants are seriously wounded. Interestingly, conflicts in a wide range of species are often resolved without dangerous escalation. In the 1950s and 1960s, many biologists believed animals refrained from lethal violence because it was good for the species. All individuals are better off, reasoning went, when conflicts are resolved without destructive violence. By the late 1960s, such explanations had been discredited because they relied on group rather than individual benefits (see Chapter 6). The old explanation for restraint in contests was gone, but there wasn't any replacement. Then in 1973 the eminent geneticist John Maynard Smith and an eccentric retired engineer named George Price[6] explained why sometimes selection favored restraint, and why sometimes it didn't. This model, called the Hawk-Dove game, illustrates the power of simple models. With one paper, Maynard Smith and Price changed the way biologists viewed animal contests.

In this chapter, you will learn about the Hawk-Dove game and some of its many descendants. We begin with the Hawk-Dove game because it provides a simple introduction to a key feature of social behavior—individual fitness typically depends on the behavior of others. In the last chapter, we analyzed a simple haploid model of viability selection that showed that the genotype with the highest fitness came to predominate in the population. Straightforward fitness maximization is a robust principle that can be used to predict evolutionary outcomes in many domains.[7] However, it doesn't work for social behavior because the best thing to do depends on what others in the population are doing. For example, whether it pays an individual to fight or flee depends upon whether its opponents are likely to fight or flee. These sorts of payoffs are *frequency dependent* because the success of each strategy depends in some way on the frequencies of the other possible strategies as well as on each strategy's own frequency. When fitness is frequency dependent, we have to use evolutionary game theory. In this chapter we introduce the basic ideas of evolutionary game theory and apply them to understand the evolution of animal contests.

2.1   The Hawk-Dove game

Imagine a species in which pairs of individuals come into conflict over an important resource such as a food cache or a territory. At any moment during the conflict, each individual can fight, display, or run away. Individuals in the population (which is very large) have an asexually inherited strategy that specifies how they behave. We begin by analyzing a model in which there are only two strategies, Hawk and Dove. Hawks always fight and never retreat. Doves display at first but run away if attacked. Thus there are three kinds of pairs: two Hawks, two Doves, and one of each. Each of these interactions entails a different outcome.

- When two Hawks meet, a fight ensues, and one wins and the other loses. The winner gets the resource, and the loser is injured. Which individual is the winner may be affected by many things, for example, size or dominance rank, but we assume that these factors are uncorrelated with anything we have knowledge of in this model. This means we can assume that the winner is chosen at random. This is just like observing a large number of boxing matches. Obviously there are reasons in each match that one or the other boxer won. Some boxers are faster, stronger, or smarter than their opponents. But if we have no information about these things, and all we know is that both contenders are boxers, then which boxer wins is random.

- When a Hawk and a Dove meet, the Hawk gets the resource, and the Dove gets nothing. The Dove also escapes uninjured.

- When two Doves meet, they display, and a random individual is eventually assigned the resource when one of them quits (or if it is divisible, they can share it).

We want to know which strategy will be favored by natural selection.

To answer this question, the first step is to write a *payoff matrix* that specifies the fitness consequences of each the four possible interactions. Suppose that the value of the resource is v, and the cost of losing a fight is c. We specify these benefits and costs as parameters rather than fixed numbers like 6 or 103 so that we can derive the consequences for any value that these parameters might take. If we used fixed numbers, we would have to rederive the evolutionary equilibria each time we used new values.

Now, let's consider each kind of interaction in turn. When two Hawks meet, each has a 50 percent chance of winning and gaining v and a 50 percent chance of losing and getting $-c$. Thus the *expected* payoff for each Hawk is $v/2 - c/2$, or $(v - c)/2$. When a Hawk meets a Dove, the Hawk receives the resource, v. The Dove receives zero, but does not incur a cost because it does not fight. Two Doves either flip a coin or share the resource so each receives $v/2$ on average. With these expressions, we get the payoff matrix shown in Table 2.1.

In this book, we indicate payoff expressions with a capital V. In this case the payoffs are

$$V(H|H) = \frac{v-c}{2} \qquad\qquad V(H|D) = v$$

$$V(D|H) = 0 \qquad\qquad V(D|D) = \frac{v}{2}.$$

Read $V(H|D)$ as "the payoff to a Hawk, given that it interacts with a Dove." The next step is to calculate what each strategy receives on average, across all interactions. Let p be the frequency of Hawks in the population. Then, assuming that individuals interact at random, we obtain for the expected fitness of a Hawk

$$W(H) = w_0 + pV(H|H) + (1-p)V(H|D)$$

$$= w_0 + p\frac{v-c}{2} + (1-p)v.$$

Payoffs in the matrix are the *changes* in fitness due the behavior in question. The parameter w_0 is the average fitness of individuals in the population, excluding the social interaction we are focusing on. Call this the *baseline fitness*. Because selection depends on differences in fitness, w_0 will vanish in most of our algebra. However, it is still very important for two reasons: First, because we interpret fitnesses as probabilities of survival (possibly times fertility), they must be positive. This means that $w_0 > c$. More important, the baseline

Table 2.1: Payoffs in the basic Hawk-Dove game. Values in each cell are fitness consequences to the player on the left (row player).

	Hawk	Dove
Hawk	$(v-c)/2$	v
Dove	0	$v/2$

fitness determines the strength of selection. When it is big compared to the effect of the behavior in question, selection changes gene frequencies slowly. When it is small, selection is strong. Biologists usually think that selection on any specific behavior is quite weak from generation to generation.[8] Moreover, when selection is strong, the sometimes unrealistic assumption of discrete, nonoverlapping generations can lead to complicated, chaotic trajectories that are probably not important in most evolutionary situations. Thus we will usually assume w_0 is large.

The expected payoff to a Dove is written similarly:

$$W(D) = w_0 + pV(D|H) + (1-p)V(D|D)$$
$$= w_0 + (1-p)\frac{v}{2}.$$

The expressions for $W(H)$ and $W(D)$ correspond to $V(A)$ and $V(B)$ in the viability selection model we began with in the last chapter. Therefore we already know that the frequency of Hawks will change each generation according to a recursion for viability selection, with $W(H)$ and $W(D)$:

$$p' = \frac{pW(H)}{\bar{w}} = \frac{pW(H)}{pW(H) + (1-p)W(D)}.$$

Subtracting p from both sides yields the difference equation:

$$\Delta p = p(1-p)\frac{W(H) - W(D)}{\bar{w}}.$$

2.1.1 Evolutionary stable strategies

This recursion tells us what happens to the frequency of Hawks and Doves over one generation. As usual, we want to know the long-run evolutionary outcome. This model is already too complex for an explicit solution, so we turn to our next best alternative, equilibrium analysis. This means we find the equilibria of the system and determine which are stable. In particular, we want to know:

1. Is any strategy stable against invasion by every other?

2. If no strategy can resist invasion by all others, is there a stable mix of these strategies?

It turns out that there are clever ways to answer these questions without doing any of the calculus from Chapter 1. Let's see how.

Imagine a population in which Doves are very rare (that is, $p \approx 1$). This means that the Hawks hardly ever interact with Doves, and therefore their average fitness is determined by interactions with other Hawks:

$$W(H) \approx w_0 + (1)V(H|H) + (1-1)V(H|D)$$
$$\approx w_0 + V(H|H)$$
$$\approx w_0 + \frac{v-c}{2}.$$

Now consider the rare mutant Dove who displays but never fights, and instead retreats if its opponent escalates. Since the chance of meeting another Dove is miniscule, its fitness is also determined by interactions with Hawks, and its fitness is

$$W(D) \approx w_0 + V(D|H) = w_0 + 0.$$

The Dove never wins the resource, but then it never gets beaten up either. When evaluating whether a strategy can invade (or resist invasion), what matters is how each type— including the common type—fares against the common type, in this case Hawks.

So in a world of all Hawks, Doves do better when $W(D) > W(H)$. Using the expressions just derived, we obtain

$$w_0 + \frac{v-c}{2} < w_0 + 0$$

or

$$v < c.$$

Doves can invade a population of Hawks when the cost of losing a fight, c, exceeds the value of the resource being fought over, v. When the cost is less than the value, it always pays to play Hawk. But when fights are costly, a rare Dove can invade the population.

Next let's determine whether a population of Doves can resist invasion by rare Hawks. The fitness of a Dove in a population of Doves is given by

$$W(D) \approx w_0 + \frac{v}{2}.$$

Doves nearly always interact with other Doves and therefore gain half the resource on average. The fitness of a rare Hawk in a population of Doves is given by

$$W(H) \approx w_0 + v.$$

The rare Hawk always interacts with a Dove and therefore always wins the resource. Since $v > v/2$, rare Hawks can always invade a population of Doves.

The results so far suggest two kinds of outcomes for the basic model of conflict. When the cost of losing a fight is smaller than the value of the resource ($c < v$), it pays on average to fight. However, when fights are costly ($c > v$), it pays to back off instead of escalating. Hence, neither Hawks nor Doves are *evolutionary stable strategies* (ESSs) when $v < c$. An ESS is a strategy that can resist invasion by any other available strategy, provided each other strategy invades alone and in small numbers.

Back to our model. When $v < c$, Doves can invade Hawks, and Hawks can likewise invade Doves. This implies that some mix of Doves and Hawks will exist in the population at equilibrium. How do we calculate this mix? From our basic difference equation we know that an equilibrium exists if

$$\Delta p = p(1 - p)\frac{W(H) - W(D)}{\bar{w}} = 0.$$

This can happen when either $p = 0$ or $p = 1$, or when $W(H) = W(D)$. When we analyzed the viability selection model in Chapter 1 this third possibility was uninteresting because fitnesses were constant. But now the fitnesses depend on the frequency of Hawks, and so there may be some frequency at which the fitnesses of Hawks and Doves are equal. Let $\hat{p}$ be the frequency of p at equilibrium. By setting $p = \hat{p}$ and $W(H) = W(D)$ and solving for $\hat{p}$, we can find the frequency of p at which there is a stable mix of Hawks and Doves:

$$w_0 + \hat{p}\frac{v-c}{2} + (1-\hat{p})v = w_0 + (1-\hat{p})\frac{v}{2}$$

$$\hat{p} = \frac{v}{c}.$$

The system will come to rest when a fraction v/c of the population are Hawks and a fraction $1 - v/c$ are Doves. If Hawks become more common, selection will favor Doves. If Doves become more common, selection will favor Hawks. An equilibrium at which more than one type persists is called a *mixed equilibrium* by game theorists, *internal equilibrium* by mathematicians, and a *polymorphic equilibrium* by population geneticists.

We can see all of the equilibria and their dynamics by plotting Δp against p (Figure 2.1). There is an equilibrium wherever $\Delta p = 0$. Equilibria exist at $\hat{p} = 1$ and $\hat{p} = 0$, but, as we have seen, neither of these equilibria is stable when $v < c$. The third equilibrium lies at $\hat{p} = v/c$. This equilibrium is stable. How do we know? We can derive the stability of either, of the first two equilibria by asking if a rare strategy has higher fitness than the common type. Then all we have to do is compare fitness expressions under these boundary conditions, as we did above to see if either Hawks or Doves can invade the other. We can't use this method for the internal equilibrium because both strategies are already present. But we know the mixed equilibrium is stable in this case because neither of the pure equilibria at $\hat{p} = 1$ and $\hat{p} = 0$ is stable. A unique internal equilibrium (at $0 < \hat{p} < 1$) is stable as long as the fitness functions don't have any kinks

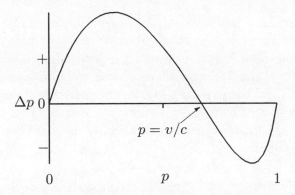

Figure 2.1: Dynamics of the Hawk-Dove game with $v = 2$ and $c = 3$. The x-axis is the frequency of Hawks, p. On the left, when Hawks are rare, the change in their frequency, Δp, is positive, and Hawks increase in frequency. On the right, Hawks are common, Δp is negative, and Doves increase in frequency. Where the dynamics crosses the x-axis, there is one stable equilibrium at $\hat{p} = 2/3$.

or discontinuities and selection is weak. If there is more than one internal equilibrium, we have to take the derivative of the recursion and evaluate it at the equilibrium, as we did in Chapter 1. The curious reader should verify that the linear approximation method tells us that the equilibrium $\hat{p} = v/c$ is stable whenever $v < c$, that is, whenever it exists.

The Hawk-Dove model has two qualitatively different outcomes depending on the ratio of v to c. When the value of the resource exceeds the cost of losing a fight, Hawks dominate. In this world, displaying never pays. When fights are costly and the resource has value ($v > 0$), a stable mix of Doves and Hawks always exists, and this stable mix is reached when the frequency of Hawks is equal to v/c.

2.1.2 Finding ESSs

We're going to find the ESS's of lots of different models. So it will be useful to summarize the basic steps. First, take each pure strategy x (Hawk or Dove, above) and ask if $V(x|x) >$ $V(y|x)$, where y is any other pure strategy. If this is true, then x can resist invasion by rare y individuals. We can then say that x is an ESS against y. If $V(y|x)$ is instead greater than $V(x|x)$, then the rare strategy can invade x, and x is therefore not an ESS. This process is justified by the simple stability conditions of pure equilibria ($\hat{p} = 1$ or $\hat{p} = 0$) we derived in Chapter 1.

Second, ask if there are any internal equilibria comprised of combinations of any of the pure strategies. These are found by asking when the fitness of any two (or more) strategies is equal. If we have two strategies x and y, then we ask if and when $W(x) = W(y)$ and solve for the frequency of x. These equilibria may be stable or unstable. If there is only one internal equilibrium and both of the pure equilibria ($\hat{p} = 1$ and $\hat{p} = 0$) are unstable, then the internal equilibrium will be stable in most cases. If instead both of the pure equilibria are stable, then the internal equilibrium is unstable. (The curious student can verify this by drawing a graph of a recursion with two stable pure equilibria.) Finally, we can always use the linear estimation technique outlined in Chapter 1 to find the local stability of the equilibrium. (So far, you only know how to do this in a one-dimensional system. In Chapter 7, you will learn how to determine the linear, aka asymptotic, stability of systems in more than one dimension.)

2.2   Retaliation

President Teddy Roosevelt famously summarized his foreign policy as "speak softly and carry a big stick." Being nice but retaliating in the face of aggression is a well-known strategy in human affairs, so maybe it is used by other animals too. It would be nice to know if the simple model we derived falls apart when we introduce a "speak softly and carry a big stick" strategy.

Consider a third strategy named *Retaliator*, which plays Dove unless its opponent plays Hawk, in which case it switches

to Hawk. Paired with a Hawk, a Retaliator acts just like a
Hawk. Paired with a Dove, a Retaliator acts just like a Dove.
It can do this because the Hawk-Dove game has two stages:
signal (fight or display) followed by behavior. When the Retaliator sees a "fight" signal, it fights back. Otherwise it
displays just as a Dove, and gets the resource half the time.
The payoff matrix is shown in Table 2.2.

Let q be the frequency of Retaliators in the population,
so that $1 - p - q$ is the frequency of Doves. Then the average
fitness of each strategy is

$$W(H) = w_0 + (p + q)\frac{v - c}{2} + (1 - p - q)v$$

$$W(D) = w_0 + (1 - p)\frac{v}{2}$$

$$W(R) = w_0 + p\frac{v - c}{2} + (1 - p)\frac{v}{2}.$$

If you feel that you can't guess what's going to happen, join
the club. The model is suddenly much more complicated, and
we're going to have to learn some new tools to understand
the system.

Let's begin by asking what happens at each of the pure
equilibrium, all Hawks, all Doves and all Retaliators. We
already know that rare Hawks can always invade a population
of Doves, and rare Doves can invade a population of Hawks

Table 2.2: Payoff matrix in Hawk-Dove-Retaliator game. Payoffs
are to the player on the left (row player).

	Hawk	Dove	Retaliator
Hawk	$\dfrac{v - c}{2}$	v	$\dfrac{v - c}{2}$
Dove	0	$v/2$	$v/2$
Retaliator	$\dfrac{v - c}{2}$	$v/2$	$v/2$

as long as $v > c$. So, neither Hawks or Doves is an ESS in the expanded game. What about Retaliators?

A population of Retaliators can resist invasion by rare Hawks when

$$
\begin{aligned}
V(R|R) &> V(H|R) \\
\frac{v}{2} &> \frac{v-c}{2} \\
c &> 0.
\end{aligned}
$$

which is always satisfied. Retaliators can resist invasion by Hawks because the Retaliators all get along, while rare Hawks pick a fight and does worse than the average. However, Retaliators cannot resist invasion by rare Doves because both have the same fitness, $w_0 + v/2$. In fact, when Hawks are absent, Retaliators and Doves always have the same fitness. Doves never fight, and Retaliators never initiate a fight, so in a population of only Doves and Retaliators there will be no escalated contests and no injuries. Everyone will get an average payoff of $v/2$. Thus, any mix of Doves and Retaliator is an equilibrium, but not an ESS. This sort of result is often called "neutral stability". Dove does not increase, but neither does Retaliator. Instead, processes we haven't modeled, like drift and mutation, will determine the mix of the two strategies.

Moreover, it seems likely that mixtures of Doves and Retaliators will eventually be invaded by Hawks. Mixtures of Doves and Retaliators are only neutrally stable so it is plausible that the frequencies of the two types will drift. We know that a population of Retaliators can never be invaded by Hawks. So it stands to reason that Hawks will be able to invade mixtures of Doves and Retaliators, if the frequency of Retaliators drifts too low. Let's determine the threshold frequency of Retaliators. In a population in which Hawks are rare, the fitness of Doves and Retaliators is:

$$
W(D) = W(R) = w_0 + \frac{v}{2}. \tag{2.1}
$$

Let q be the frequency of Retaliators. The fitness of a rare invading Hawk is:

$$W(H) = w_0 + q\frac{v-c}{2} + (1-q)v. \qquad (2.2)$$

Assuming again that $v < c$, the fitness of Hawks increases as the frequency of Retaliators decreases. Hawks will be able to invade a population of Doves and Retaliators when q, the frequency of Retaliators, is such that the fitness of a rare Hawk (Expression 2.2) exceeds the expected fitness of a Dove or Retaliator (Expression 2.1):

$$q\frac{v-c}{2} + (1-q)v > \frac{v}{2}.$$

Solving for q:

$$q < \frac{v}{v+c}.$$

Above this threshold, rare Hawks encounter too many Retaliators and do worse than Doves. Hence they cannot invade. Below this frequency, Hawks are able to exploit enough Doves to make the fights with Retaliators worth it, and they invade. The trouble with speaking softly and carrying a big stick is that it allows Doves to thrive and, because these Doves speak softly but don't carry any stick, they are food for Hawks.

Where does the system go once Hawks invade? We can answer this question by asking whether Retaliators can invade the mix of Doves and Hawks we derived in the model without Retaliators. Recall that there was a stable mix of Doves and Hawks when $v < c$, at $\hat{p} = v/c$. Let's see if Retaliators invade this equilibrium.

At the mixed Hawk/Dove equilibrium, $q = 0$ and $p = v/c$. With these assumptions, the fitness of Hawks and Doves is:

$$W(H) = W(D) = w_0 + \left(1 - \frac{v}{c}\right)\frac{v}{2}.$$

The fitness of a rare Retaliator under the same conditions:

$$W(R) = w_0 + \frac{v}{c}\frac{v-c}{2} + \left(1 - \frac{v}{c}\right)\frac{v}{2}.$$

Thus a rare Retaliator can invade the mixed Hawk/Dove equilibrium when

$$W(R) - W(H) > \frac{v}{c}\frac{v-c}{2} > 0$$

which is true only when $v > c$. But if $v > c$, the mixed Hawk/Dove equilibrium does not exist, and so this mixed equilibrium is stable against invasion by rare Retaliators whenever it is present. Thus, the most plausible long run outcome is the same mixture of Hawks and Doves as in the simpler game. Adding Retaliators doesn't really matter.

As an aside, if Retaliators had been able to invade, we would have now gone on to find any possible mixed equilibria of all three strategies, and this is done just as you'd expect, by asking what values of p and q make $W(H) = W(D) = W(R)$.

2.2.1   Diagramming a three-strategy system

Here it is helpful to introduce a useful way to diagram systems like the Hawk-Dove-Retaliator model that have three possible strategies. What we want is a phase diagram, that is, a plot that shows the direction the system takes at any given mix of Hawks, Doves, and Retaliators. We could plot this phase diagram on a regular rectangular grid with one strategy on the x-axis and one on the y-axis. The drawback of this is that the rectangular plot privileges two of the strategies over the third. We want a more unbiased diagram.   To plot mixes of three things, game theoreticians (as well as chemists) often use a *ternary plot*. Ternary plots are also called de Finetti diagrams, barycentric plots, and mixture plots. The plot is an equilateral triangle (Figure 2.2). Each vertex is a strategy. If a point is at a vertex, the system at that point contains only a single strategy. In the center of the triangle lies a point which represents the system when each strategy is present in equal proportions. Thus the frequency of any strategy is given by the length of the line segment perpendicular to the face opposite the strategy's vertex, extending from the face to the point.

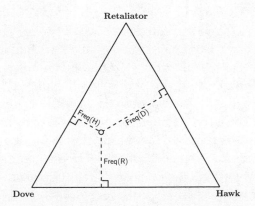

Figure 2.2: Ternary plot for the Hawk-Dove-Retaliator game. The internal point represents a specific mix of each strategy. The frequency of each strategy is given by the length of the line segment perpendicular to the face opposite the strategy's vertex, extending from the opposite face to the point.

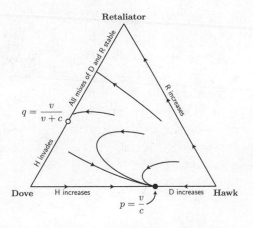

Figure 2.3: Ternary plot of the Hawk-Dove-Retaliator system dynamics when $v/c = 2/3$. Arrows show evolutionary trajectories originating from populations at different initial mixes of Hawks, Doves, and Retaliators. Above a certain frequency, Retaliator excludes Hawk. However, Retaliators cannot invade a population of Hawks and Doves.

Figure 2.3 plots the phase diagram for the game we derived above. The line segment between Hawks and Doves is the basic Hawk-Dove game. Each curve within the diagram shows how the system evolves from a given initial mix of the three strategies. Note the stable equilibrium at $\hat{p} = v/c$ (in this case, 2/3), which we derived at the beginning of this chapter. Along the segment connecting Doves and Retaliators, there is a point below which Hawks invade and above which Hawks are kept out but selection is neutral with respect to Doves and Retaliators. This is the threshold value of q we derived earlier: $q < v/(v + c)$.

2.3   Continuous stable strategies

Sometimes there are parameters in a model which describe some aspect of a strategy. For example, we could define a strategy which plays Hawk a proportion h of the time and Dove the rest. Obviously there is a family of such strategies, comprised of all the values of h from zero to one. It would be useful to know what value of h natural selection would prefer. Here we will take a short detour to learn how to solve such *continuous stable strategy* (CSS) problems.

Recall the basic Hawk-Dove game with $v < c$. Instead of having two pure strategies, Hawk and Dove, this time we assume a family of strategies which play Hawk some proportion h of the time and Dove the rest of the time. Let's call this strategy HD_h for a given value h. The payoff to such a strategy against itself is

$$V(HD_h|HD_h) = h \left\{ \underbrace{\underbrace{h\frac{v-c}{2}}_{\text{opponent plays Hawk}} + \underbrace{(1-h)v}_{\text{opponent plays Dove}}}_{\text{payoff when plays Hawk}} \right\}$$

$$+ (1-h) \underbrace{\left\{ h(0) + (1-h)\frac{v}{2} \right\}}_{\text{payoff when plays Dove}}.$$

Now imagine a strategy $HD_{h+\delta}$ which plays Hawk a proportion $h + \delta$ of the time, where δ is any small real number,
positive or negative. We want to know when $HD_{h+\delta}$ will invade HD_h. The gambit here is to imagine that mutations
result only in small changes to the common strategy HD_h, so
we only have to consider invasion by strategies like $HD_{h+\delta}$ in
the immediate neighborhood of the common type. $HD_{h+\delta}$'s
payoff against HD_h is

$$V(HD_{h+\delta}|HD_h) = (h + \delta)\left\{ h\frac{v - c}{2} + (1 - h)v \right\}$$
$$+ (1 - h - \delta)\left\{ h(0) + (1 - h)\frac{v}{2} \right\}.$$

In order to find out when $HD_{h+\delta}$ will invade HD_h, we ask
when
$$V(HD_{h+\delta}|HD_h) > V(HD_h|HD_h)$$

and solve for h. Essentially, we are asking what values of h
allow invasion by different values of h.

In Box 2.1 we show that the only un-invadable value of h
is

$$h^* = \frac{v}{c}. \tag{2.3}$$

This result mirrors our previous mixed equilibrium of pure
strategies (Hawk and Dove), but for mixed strategies which
play Hawk and Dove with different probabilities. Only a
strategy which plays Hawk v/c of the time is a stable strategy.

The same methods can be used to derive CSS results for
many other models. Often, some of the parameters in our
models are plausibly at least partly controlled by an evolved
psychology. In these cases, CSS analyses can provide insights
into what natural selection would do with those parameters
if it had its way. Of course, natural selection may not always
get its way, but that is true of any of these models.

The topic of pure versus mixed strategies has a literature
of its own which we won't go into here. It is worth mention-
ing, however, that the solution with the mixed strategy and

Box 2.1 Deriving the CSS value of h

We're going to take a shortcut here and write down the difference in payoff to the two strategies term by term. This is the same as writing one payoff on one side of the inequality and the other on the other side and then subtracting and factoring common terms, but it saves us some steps. HD_h can be invaded by $HD_{h+\delta}$ when

$$\frac{v-c}{2}\{h^2 - (h+\delta)h\} \quad + \quad v\{h(1-h) - (h+\delta)(1-h)\}$$
$$+ \quad \frac{v}{2}\{(1-h)^2 - (1-h-\delta)(1-h)\}0.$$

Simplifying

$$\frac{v-c}{2}\{-\delta h\} + v\{-\delta(1-h)\} + \frac{v}{2}\{\delta(1-h)\} < 0.$$

At this point we can divide both sides by δ, since it is in every term. However, keep in mind that δ can be either a positive or a negative number, so we have to keep track of the signs now. Assuming $\delta > 0$ and multiplying everything by 2, we obtain

$$(v - c)\{-h\} + 2v\{-(1 - h)\} + v(1 - h) < 0$$
$$h < \frac{v}{c}.$$

If $h < v/c$, a mutant who plays Hawk *more often* ($\delta > 0$) does better. This result should already make plenty of sense. We derived the mixed equilibrium for Hawks and Doves and found that if less than a proportion v/c of the population is Hawks, Doves do worse than Hawks. The result above is the same for the mixed strategy HD_h. If $h < v/c$, any mutant who plays Hawk a bit more often does better. Now, assuming $\delta < 0$, we obtain

$$(v - c)\{-h\} + 2v\{-(1 - h)\} + v(1 - h) > 0$$
$$h > \frac{v}{c}.$$

The sign just gets reversed, of course. A mutant who plays Hawk a little *less often* will invade only when $h > v/c$. If the common type is playing Hawk too often, a more peaceful mutant invades. Thus we know that the CSS value of h—let's call this $h^\star$—is Expression 2.3 in the main text.

the polymorphic equilibrium in a model with pure strategies turned out to be the same because the payoffs in the Hawk-Dove game are *symmetrical*. That is, each player faces the same payoff matrix. The payoff matrixes we have written so far look the same whether we look at them from the perspective of the row player or the column player. When this is not true, it matters whether the strategies are themselves mixed or not.

2.4 Ownership, an asymmetry

So far, we have only studied games in which both players are identical except for their strategy. They arrive at the same time, they are the same size, they each have the same chance of winning, etc. It would be useful to know how animals might deal with situations in which they are different in some way.

First, let's consider asymmetries which are observable but otherwise don't affect the chance of winning the contest (or more generally the outcome of the social interaction) or the value of the resource. While such *uncorrelated* asymmetries do not directly affect who wins the contest, they may still strongly affect the outcome.

Consider a situation in which one individual arrives before the other and claims a valued resource. We will consider a strategy called Bourgeois which plays Hawk when it has arrived first (is the "owner" of the resource) but plays Dove when it is the intruder. (John Maynard Smith is responsible for the sly political jibe implied by the strategy's name.) When two Bourgeois individuals meet, one agrees to leave, and the other claims the entire resource. We assume that each player has an equal chance of being the owner of the resource in any given generation. That is, being an owner is uncorrelated with strategy—being Bourgeois does not make the individual any more likely to be the owner. It just changes how the individual acts when it is the owner.

The payoff to a Bourgeois when it meets another Bourgeois is

$$V(B|B) = \underbrace{\frac{1}{2}v}_{\text{arrives first}} + \underbrace{\frac{1}{2}(0)}_{\text{arrives second}} = \frac{v}{2}.$$

What happens when Bourgeois meets a Hawk or Dove? If a Bourgeois meets a Hawk, half the time the Hawk is owner and the Bourgeois goes away, letting the Hawk win. The other half of the time, the Hawk is intruder, and so the Bourgeois fights, leading to an escalated contest. Therefore

$$V(B|H) = \frac{1}{2}\frac{v-c}{2} + \frac{1}{2}(0).$$

The payoff to a Hawk who meets a Bourgeois is similarly derived:

$$V(H|B) = \frac{1}{2}v + \frac{1}{2}\frac{v-c}{2}.$$

A Dove will end up sharing the resource with the Bourgeois half the time and giving up the other half of the time:

$$V(D|B) = \frac{1}{2}\frac{v}{2} + \frac{1}{2}(0)$$

and

$$V(B|D) = \frac{1}{2}v + \frac{1}{2}\frac{v}{2}.$$

Once again we ask what happens at each of the pure equilibria, and then whether Bourgeois can invade the mixed Hawk-Dove equilibrium. Suppose first that Hawks are common. We know that Doves can invade, when $v < c$. Bourgeois can invade when

$$V(B|H) > V(H|H)$$

$$\frac{v-c}{4} > \frac{v-c}{2}.$$

Thus both Dove and Bourgeois can invade a population of Hawks (again assuming $v < c$). When Doves are common, Bourgeois can invade when

$$V(B|D) > V(D|D)$$

$$\frac{3v}{4} > \frac{v}{2}.$$

And this is always true. Thus both Hawks and Bourgeois can invade the pure Dove equilibrium. What happens then at the pure Bourgeois equilibrium? Bourgeois is an ESS when it has higher fitness than both Hawks and Doves:

$$V(B|B) > V(H|B)$$

$$\frac{v}{2} > \frac{v}{2} + \frac{v-c}{4}$$

$$v < c.$$

So Bourgeois can resist invasion by Hawks. What about Doves? The condition is

$$V(B|B) > V(D|B)$$

$$\frac{v}{2} > \frac{v}{4}.$$

Bourgeois has higher fitness than both Hawks and Doves. Neither can invade the pure Bourgeois equilibrium. Since Bourgeois can invade both other pure equilibria, there is really no need to check the mixed Hawk/Dove equilibrium. It is easy enough to prove, however, that Bourgeois can invade that equilibrium as well. We already know the fitness of either a Hawk or a Dove at the mixed Hawk-Dove equilibrium is

$$W(H) = W(D) = w_0 + \left(1 - \frac{v}{c}\right)\frac{v}{2}.$$

An invading Bourgeois will have fitness

$$W(B) = w_0 + \frac{v}{c}\frac{v-c}{4} + \left(1 - \frac{v}{c}\right)\frac{3v}{4}$$

$$= w_0 + \left(1 - \frac{v}{c}\right)\frac{v}{2}.$$

A rare Bourgeois has the same fitness as a Hawk or Dove and can drift in. As Bourgeois becomes more common, relative fitnesses will be determined by how each does against Bourgeois. Bourgeois against itself receives a higher payoff than both Hawk and Dove against Bourgeois, and so selection will increase the frequency of Bourgeois until the population reaches the pure Bourgeois equilibrium. There is only one truly stable equilibrium in this system. Bourgeois does well against either pure strategy because it is able to establish a convention for efficiently dividing the resource across generations. Once common, this convention punishes any strategy that does not follow it.

One interesting aspect of this model is that there is an anti-Bourgeois strategy, call it Anarchist ("Property is theft" was the slogan of the famous nineteenth-century anarchist Proudon), which plays Dove when owner and Hawk when intruder. This strategy will also be an ESS, for the same reasons Bourgeois is an ESS. And either Bourgeois or Anarchist will be able to exclude the other, once common. However, since an Anarchist would be displaced by another, or even by the one it had just displaced, this strategy seems unlikely to be stable in even a slightly more realistic model.

2.5   Resource holding power

The asymmetry we examined above, ownership, is uncorrelated with the ability to win contests. Now let's explore a *correlated* asymmetry, such as fighting ability, that is associated with the ability to win contests. In many cases, animals differ noticeably in size, strength, or age, and these differences are predictive of the outcomes of contests. These asymmetries are often referred to as resource holding power (RHP).[9] It stands to reason that animals might use these asymmetries to resolve contests.

Let's consider a strategy called Assessor. Assessor plays Hawk when it is larger (or stronger, older, smarter, etc.) than its opponent and plays Dove when it is smaller than its opponent. Let x be the chance of an individual winning when it is larger than its opponent, and assume that $x > 1/2$. Thus the chance of winning when an individual is smaller than its opponent is $(1 - x) < 1/2$.

Assume that each individual in a pair has an equal chance of being larger than its opponent. The payoff to an Assessor facing another Assessor is then

$$V(A|A) = \underbrace{\frac{1}{2}v}_{\text{bigger}} + \underbrace{\frac{1}{2}(0)}_{\text{smaller}} = \frac{v}{2}.$$

Half the time, the focal individual is larger and receives v. The other half of the time, it is smaller and receives 0. The other payoffs are constructed via the same logic. For an Assessor versus a Hawk,

$$V(A|H) = \underbrace{\frac{1}{2}\left(xv - (1-x)c\right)}_{\text{bigger}} + \underbrace{\frac{1}{2}(0)}_{\text{smaller}} = \frac{xv - (1-x)c}{2},$$

$$V(H|A) = \underbrace{\frac{1}{2}v}_{\text{bigger}} + \underbrace{\frac{1}{2}\left((1-x)v - xc\right)}_{\text{smaller}}.$$

For an Assessor versus a Dove,

$$V(A|D) = \underbrace{\frac{1}{2}v}_{\text{bigger}} + \underbrace{\frac{1}{2}\frac{v}{2}}_{\text{smaller}} = \frac{3v}{4},$$

$$V(D|A) = \underbrace{\frac{1}{2}\frac{v}{2}}_{\text{bigger}} + \underbrace{\frac{1}{2}(0)}_{\text{smaller}} = \frac{v}{4}.$$

Assessor clearly is an ESS against Dove ($V(A|A) > V(D|A)$). What about against Hawk? Assessor resists invasion by rare Hawks when

$$\frac{v}{2} > \frac{v}{2} + \frac{(1-x)v - xc}{2}.$$

This simplifies to

$$xc > (1-x)v.$$

Consider the extreme case of $x = 1$, such that the larger individual always wins. Then Assessor is an ESS as long as $c > 0$. In fact, v can be greater than c, and Assessor is still an

ESS. When x is very small, Assessor can be an ESS against Hawk as long as c is big enough. That is, if losing is very costly, it pays to pick your fights wisely. In the case $x > 1/2$, Assessor is an ESS whenever $v < c$. Hawk is an ESS against Assessor $(V(H|H) > V(A|H))$ when the opposite is true:

$$(1 - x)v > xc.$$

So even if $v > c$, a condition in which Hawk dominates Dove, Assessor can sometimes invade and exclude Hawk. Suppose, for example, that $x = 3/4$, $v = 2$, and $c = 1$. Then Hawk is not an ESS, but Assessor is. The lesson here is that it really pays to choose your fights wisely. If individuals pick only fights they are likely to win, universal aggression can be stable, even when the costs of losing a fight are very large.

An obvious question is what happens when individuals can use different asymmetries, such as ownership and size, simultaneously. You will answer this question when you work problem 4 at the end of this chapter.

2.6   Sequential play

So far we have examined games in which both players act simultaneously, or at least each player acts before it knows what its opponent will do. Now, let's consider a sequential game in which one player announces and is committed to its behavior before the second player chooses its behavior. Sequential games are often analyzed using game trees (economists call these *extensive-form*, as opposed to *strategic-form*, games). The tree in Figure 2.4(a) represents the original simultaneous Hawk-Dove game. The first player chooses Hawk or Dove, then the second player chooses Hawk or Dove. The dashed outline enclosing the two decision nodes for Player 2 is called the information set. That is, Player 2 does not know which of these two nodes she is at.

Removing the information set makes the game sequential, as in Figure 2.4(b). Now, provided that Player 1 cannot

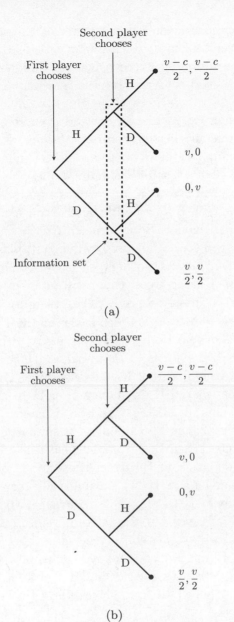

(a)

(b)

Figure 2.4: Sequential play in the Hawk-Dove game. (a) Extensive-form representation of the usual simultaneous game. The dashed outline represents the information set. (b) Extensive-form representation of the sequential game.

change its mind, Player 2 can be coerced into playing Dove against Player 1's Hawk. To see why, let's learn how to find ESS strategies in game trees using a technique called *backwards induction* that is widely used in economics. We could find the ESS strategy here the usual way, by writing a fitness expression for each strategy and then comparing these fitnesses to one another in a pairwise fashion. However, consider how many strategies there are in this game already. A full strategy must say what the animal does when it is Player 1 and Player 2. Thus there are at least four strategies: HH, HD, DH, and DD. But none of these is contingent upon the other player's behavior. So we'd need to write expressions for every strategy of the sort "play D when first; play H if second and first player plays D, otherwise play D." For even moderately complex game trees, there will be many strategies in the possible set, and ESS analysis can be quite a chore.

Backwards induction is much easier and works as follows. Begin with the final nodes of the tree. At each node, ask what the ESS behavior is *at that node*. Take the upper-right node in the tree in Figure 2.4(b), for example. Given that the first player has played H, is playing H or D an ESS? Clearly, playing D in this situation is the ESS behavior. In the other node for Player 2, the ESS behavior is similarly H, since the first player has chosen D. Now we repeat this process for the nodes one step back in the tree, those for Player 1 in this case. If Player 1 plays H, we have already seen that Player 2's ESS behavior is to play D, so Player 1 would receive v in this case. If Player 1 plays D, we have seen that Player 2 will play H, so Player 1 would receive nothing in this case. Since $v > 0$, Player 1's ESS behavior is to play H. Thus the complete ESS strategy says "Play H if first player. When second player, play D if first player plays H. Play H if first player plays D."

Order matters here. Player 1 forces Player 2 to cave in. There are other games in which the first player usually gets her way, as in the classic battle of the sexes game that you will analyze in this chapter's problem set.

2.6.1 Commitment and signaling

It stands to reason that if Player 2 in the game above could show Player 1 that she is committed to play H, Player 1 would be coerced into playing D. We will explore this possibility in Chapter 5.

Guide to the Literature

Evolutionary stable strategies. The term and basic definition is due to John Maynard Smith and George Price 1973. **Hawk-Dove game.** The short 1973 Maynard Smith and Price paper on Hawk-Dove includes more strategies than are normally discussed. We recommend reading it, although Maynard Smith's text *Evolution and the Theory of Games* 1982 is still the best introduction to simple models of conflict. It includes analysis of an interesting game, the "war of attrition," that we have not mentioned in the chapter. See also Bishop and Cannings 1978. Parker and Thompson 1980 applied the war of attrition to dung fly behavior. **Asymmetries.** Parker 1974 and Maynard Smith and Parker 1976 wrote the first analyses of size asymmetry in these contests. The Hawk-Dove game has been used frequently to analyze the evolution of "property rights." Grafen 1987 argued that respect for ownership asymmetries may not be as likely as the simple models suggest. Interesting papers by Susan Riechert 1984, 1986 and Reichert and Hammerstein 1983 test which asymmetry, ownership or size, used by territorial spiders. Crowley's 2000, 2003 papers address how various aspects of knowledge and discrimination of cues lead to patterns of behavioral variation. **Sequential play.** Reinhard Selten won the Nobel prize for his work on sub-game perfect equilibria in sequential (extensive form) games. See Selten 1975. The concept continues to be just as important in evolutionary models as in economic models. Several models of sequential and state-dependent Hawk-Dove games exist (see Enquist and Leimar 1983 and Houston and McNamara 1988).

Problems

1. Conflict as cooperation. It is possible to use the Hawk-Dove game to study the evolution of certain behaviors we commonly think of as cooperation. One form of the game might be called the "doing the dishes" game. In many tasks, two individuals both have a strong interest in getting the task done, but each would prefer the other individual do it. Nest-building, cooperative defense, and parental care may be examples of this kind of problem. Suppose that each of two individuals can choose to wash the dishes or not. Each player gains b when the dishes get done and must pay k if it washes alone or half that if both players cooperate to wash the dishes. The payoffs in this game are then (payoffs to player on the left) as follows:

	Rest	Wash
Rest	0	b
Wash	$b - k$	$b - \frac{k}{2}$

There are only two strategies (for the moment): *Rest*, which never washes the dishes, and *Wash*, which always washes the dishes, either alone or with the other player. Assume $b > k$, so that it always pays someone to do the dishes alone. (a) Show that this game has the same order of payoffs as the Hawk-Dove game,

$$V(H|H) < V(D|H) < V(D|D) < V(H|D),$$

when $v < c$, assuming that Wash = Dove (cooperative strategy) and Rest = Hawk (uncooperative strategy). (b) Show that there is a stable internal equilibrium at $\hat{p} = k/(2b - k)$, where p is the frequency of Rest in the population. (c) Imagine now a third strategy called Spiteful, which only washes when the other individual also washes. Spiteful always begins by offering to wash, such that two Spiteful individuals both wash when they meet, but refuses to wash if the opponent does not also wash. Show that Spiteful is an ESS against Rest when $b > k$ but never against Wash. Also show that Spiteful cannot invade the mixed Wash/Rest equilibrium you found above unless $k < 0$. Compare the evolutionary dynamics of Spiteful in this game to those of Retaliator in the Hawk-Dove game.

2. Display costs. The version of the Hawk-Dove game presented above assumes that two Doves can resolve the contest at zero cost and thereby capture the entire value of the resource between them. Since Doves display at each other, this assumption doesn't seem very plausible—the cost of displaying must dissipate some the value of the resource. Consider a more general version of the Hawk-Dove game in which Doves pay a display cost $d/2$ when they interact. Assume that $0 \le d \le v$, so that when $d = 0$, we have the original Hawk-Dove game, and when $d = v$, we have a more pessimistic version, which assumes that selection increases the amount of displaying until it consumes all of the value of the resource. (a) Write down the game matrix and prove that neither Hawk nor Dove is an ESS. (b) Show that there is a mixed equilibrium a which the frequency of Hawks is

$$\hat{q} = \frac{v + d}{c + d}.$$

(c) Does adding a display cost change the biological conclusions that we draw from the game?

3. More display costs. Add the Bourgeois strategy to the Hawk-Dove game as modified in Problem 2, but now assume that $d = v$, so that two Doves get nothing when they interact. (a) Write down the game matrix and show that

Bourgeois is an ESS. (b) Show that rare Bourgeois individuals have the same fitness as Hawks and Doves at the mixed equilibrium you derived in Problem 2. (c) Does adding a display cost change the biological conclusions that we draw from the game?

4. Correlated versus uncorrelated asymmetries. We have already examined the effects of correlated (e.g. strength) and uncorrelated (e.g., ownership) asymmetries in the Hawk-Dove game. Peter Hammerstein (1981) explored what happens when different asymmetries of different types are possible. Here you will prove that precedence can have important effects in determining which asymmetry wins.

Recall the Bourgeois strategy. Animals are randomly owners or invaders. Bourgeois plays Hawk when it is owner and Dove when it is invader. Now also recall Assessor, which plays Hawk when it is stronger than its opponent and plays Dove when it is weaker than its opponent. There is always a 50-50 chance of having greater strength than one's opponent.

(a) We already derived the conditions for Bourgeois and Assessor to be ESSs against both Hawk and Dove. Assume these conditions hold. Now show that Assessor is an ESS against Bourgeois when

$$x > \frac{v}{2(v+c)}.$$

Show also that Bourgeois is an ESS against Assessor, provided that

$$x < \frac{v+2c}{2(v+c)}.$$

(b) Assume both conditions above are met. Let p be the frequency of Assessor in a mixed population. Show that there is an unstable equilibrium at

$$\tilde{p} = \frac{v(1-2x) + 2c(1-x)}{2c}.$$

(c) Assuming $x > 0.5$, which of the two strategies is more likely to evolve? Why is this?

5. Errors in behavior. When we analyzed the Hawk-Dove-Retaliator game, we found that when Hawks were absent, Doves and Retaliators had exactly the same fitness. This allowed Doves to drift neutrally into Retaliators, which could eventually lead to Hawks invading and removing Retaliators from the population entirely. Here you will see what happens when we add errors in behavior, such that Retaliators sometimes mistake Doves for Hawks and escalate. (a) Modify the Hawk-Dove-Retaliator game assuming that a Retaliator has a chance e of mistaking an opponent playing Dove for an opponent playing Hawk. Assume that e is small, such that terms of order e^2 and higher are approximately zero. Show that the payoffs in a world with only Doves and Retaliators become

$$V(D|R) = (1 - e)\frac{v}{2},$$

$$V(R|D) = (1 + e)\frac{v}{2},$$

$$V(R|R) = (1 - 2e)\frac{v}{2} + e(v - c).$$

(b) Show that Retaliator can invade Dove when $e > 0$. (c) Show that Dove can invade Retaliator when $e > 0$ and

$$\frac{v}{2c} < 1.$$

(d) Provided both conditions above hold, let q be the frequency of Retaliator and show that there is a stable mix of Dove and Retaliator at

$$\hat{q} = \frac{v}{2c}.$$

(e) Finally, show that Hawks can invade this equilibrium provided

$$e > 1 - \frac{c}{v}.$$

How do you interpret this result? (*Hint*: Wait until the very end of the algebra to substitute in the value of $\hat{q}$.)

6. Battle of the sexes. A well-known game in economics is called "The battle of the sexes." In this game, two players, named Pat and Chris, both want to do something together in the evening, but each has different preferences. Pat wants to go to the opera. Chris wants to go to the ballgame. Both would rather do something together, however, than do either activity alone. The payoff matrix for this game is as follows:

		Chris	
		Opera	Game
Pat	Opera	2, 1	0, 0
	Game	0, 0	1, 2

(The first payoff in each cell is Pat's. The second is Chris'.)

(a) Draw the game tree for this game, assuming simultaneous play. (b) Draw the game tree for this game, assuming that Pat goes first. Find the ESS strategy for each player, using backwards induction.

Notes

[6]Maynard Smith and Price 1973.

[7]Eshel and Feldman 1984, Hammerstein 1996.

[8]There are exceptions: See John Endler's 1986 classic, *Natural Selection in the Wild*, for a survey.

[9]Parker 1974, Maynard Smith and Parker 1976.

Box 2.2 Symbols used in Chapter 2

v Value of a contested resource, in fitness units

c Cost of losing an escalated fight, in fitness units

w_0 Baseline fitness, average fitness in the absence of the domain of behavior being modeled

h Proportion of time a mixed strategy HD_h plays like a Hawk

x Probability that the larger individual wins an escalated fight

Chapter 3

Altruism and Inclusive Fitness

Life as a ground squirrel must be terrifying. Every shadow that passes overhead could mean you are doomed to be lunch for a hawk; every rustle in the bushes, dinner for a coyote. Oddly enough, ground squirrels don't always dive for safety at the first sign of a predator, but instead sometimes they stand stiffly and give a shrill alarm call that alerts other squirrels to the predator. Paul Sherman has shown that Belding ground squirrels in the western United States who give such calls are more likely to be the victims of predation than squirrels who do not.[10] Why doesn't the vigilant squirrel just duck for cover at the first sign of danger, and leave the other squirrels as raptor bait?

Evolutionary biologists refer to behaviors like alarm calls that increase the fitness of the recipients but lower the fitness of the actor as *altruism*. Squirrels who hear the call are better able to avoid the predator, but callers place themselves at risk. If genes spread because they enhance the survival and reproduction of their carriers, how can behavior of this kind evolve by natural selection? In 1975, E. O. Wilson called altruism the central theoretical problem of sociobiology.[11] Thirty years later, *Science* magazine ranked the

71

evolution of cooperation, by which they mostly meant altruism, among the "top 25 big questions facing science over the next quarter-century."[12] The evolution of altruistic behavior had a strong role in defining sociobiology, and it continues to be one of its core problems.

In this chapter, we will see why natural selection can favor altruism when relatives interact. We will also introduce several new tools for building and solving analytical models, including George Price's covariance genetics framework. In subsequent chapters we'll return to altruism several times. In Chapter 4, we'll discuss how repeated interactions and reciprocity can lead to the evolution of altruism. In Chapter 5, we'll explore how altruistic behavior can be a signal, and finally in Chapter 6, we'll see how selection within and between groups relates to the evolution of altruism.

3.1   The prisoner's dilemma

Many readers will have already heard of the *prisoner's dilemma* (or PD for short). While there are other games that produce social dilemmas, the prisoner's dilemma is the purest expression of the conflict between individual and group interests. For this reason, it continues to be a focus of interest in both biology and the social sciences.

In this book, we use a slightly stronger definition of the dilemma than is typical. Let's start with the two individual version of the prisoner's dilemma. There are two behaviors, C for cooperate and D for defect. To be a PD, the payoffs have to satisfy the two conditions, First:

$$2V(C|C) > V(C|D) + V(D|C) > 2V(D|D).$$

This expression says that cooperation increases the average fitness of the pair of individuals interacting. Two cooperators are better off on average than a cooperator and a defector, who in turn are better off than two defectors. Thus mutual cooperation maximizes group welfare. The second condition is

$$V(D|C) > V(C|C) > V(D|D) > V(C|D).$$

Table 3.1: Payoff matrix for the additive form of the prisoner's dilemma. Payoffs are to player 1.

| | Player 2 | |
Player 1	Cooperate (C)	Defect (D)
Cooperate	$b - c$	$-c$
Defect	b	0

This expression says that no matter what your partner or opponent does, you are better off defecting. These two condition show why the prisoner's dilemma *is* altruism in the sense that W. D. Hamilton[13] and Robert Trivers[14] defined it—the PD describes a situation in which there is a behavior that is mutually beneficial but individually deleterious.

In this book we will mainly focus on a special version of the PD, the additive model used by both Hamilton and Trivers. Imagine a situation in which each of two individuals can perform a behavior which increases the other's fitness by b while reducing its own fitness by c. These fitness units could be expected numbers of offspring or chances of survival. Cooperators perform the behavior and defectors do not. The payoff matrix for this game is shown in Table 3.1. If both individuals cooperate, each receives $b - c$, which maximizes the group (pair) fitness. However, if either individual switches to defection, it receives b, which maximizes individual fitness.

The PD is a true dilemma because no matter what one's opponent does, it is always better to defect. This means defection is the only ESS if, as we have assumed so far, interacting pairs are formed at random. To see why, consider two pure strategies, ALLC and ALLD. ALLC always cooperates. ALLD always defects. Assume pairs of individuals are sampled at random from a population in which ALLC exists with frequency p. The fitness of ALLC is then

$$W(\text{ALLC}) = w_0 + p(b - c) + (1 - p)(-c) = w_0 + pb - c.$$

The fitness of ALLD is similarly

$$W(\text{ALLD}) = w_0 + pb + (1 - p)(0) = w_0 + pb.$$

Since $W(\text{ALLC}) < W(\text{ALLD})$, at all values of p, ALLD will always invade and replace ALLC. Yet altruism seems to exist in nature. How can it evolve?

3.1.1   Definitions of altruism: absolute and relative

Before we start in earnest, we need to talk about definitions. We just defined altruism as a behavior which reduces the absolute fitness of the actor and increases the absolute fitness of the recipient(s). Not everyone defines altruism in this way. D. S. Wilson, among others, defines altruism as a behavior which reduces the fitness of the actor *relative* to the recipient(s). It makes a big difference which definition we use.

Absolute altruism is favored by selection in a *much* narrower range of circumstances than relative altruism. To see why, consider the following model of a shared public good. Think about an aquatic creature, say a salamander, who lives in a pond with other salamanders, and consider a behavior like defecating in the common pond. Walking a few meters away from the pond to defecate is a private cost for the actor.

Table 3.2: Payoff matrix for the common-pond example. Payoffs are to player 1.

	Player 2	
Player 1	Cooperate	Defect
Cooperate	$B - c$	$B/2 - c$
Defect	$B/2$	0

All individuals who share the pond, however, share in the benefit of cleaner water. The payoff matrix for this game is shown in Table 3.2. Cooperation means walking away from the pond, defection means defecating in the pond, B is the total group benefit, and c is the private cost of walking away from the pond (energy loss, increased risk of predation, etc.). For this game to be a PD, c must be greater than $B/2$, the marginal benefit to the actor. Otherwise $V(C|D)$ would be greater than zero, $V(D|D)$.

Suppose instead that $B/2 > c$, such that the private benefit to the actor is greater than its cost of performing the behavior (its absolute fitness increases as a result of the behavior). This might happen if the benefit of the individual's share of clean water gained by defecating outside the pond exceeds the risks incurred by walking away from the pond. If this is so, since $V(C|D) > 0$ and $V(C|C) > V(D|C)$, it always pays to cooperate, even when the other individual defects. This is not a PD. The dominant move here is to cooperate, no matter what your opponent does! Nonetheless cooperators either have the same fitness as, or lower fitness than, their pond mates—cooperation is altruism by the relative fitness definition.

Thus explanations for the evolution of altruism—and consequently how common it should be in nature—are quite different depending upon which definition of altruism one uses. The commonsense English definition of altruism includes many behaviors which are not prisoner's dilemmas.[15] However, many others are clearly PDs. Insisting that all "cooperative" behavior in nature conforms to a PD does violence to the usual meaning of the word.[16] But so is confusing the results of models which address the PD with those which address other sorts of behaviors which we often call "cooperative."

In this book we will use the absolute definition of altruism. We think that the empirical evidence demonstrates that altruism defined in this way is usually limited to close relatives. Aside from the social insects, naked mole rats, and humans, costly acts that benefit large groups are especially rare. If there were some simple solution, as some of the relative altruism models suggest, then cooperation in large groups would be much more common. Furthermore, where cooperation in large groups does exist, it is plausibly a prisoner's dilemma. The prisoner's dilemma is important.

3.2   Positive assortment

The key to understanding the evolution of altruism is nonrandom interaction. If altruistic strategies are more likely to be paired with other altruistic strategies, altruism can evolve. If interaction is random, it cannot. Let's see why.

So far, we have assumed random group formation. Pairs of individuals were sampled at random from the population such that the chance of drawing an individual of a particular strategy was just the frequency of that strategy in the population as a whole. To allow for nonrandom interaction, define the probability that an altruist (A) is paired with another altruist as $\Pr(A|A)$. Likewise, $\Pr(N|A)$ is the probability that the individual is paired with a nonaltruist, given it is an altruist. $\Pr(N|N)$ and $\Pr(A|N)$ are the probabilities that a nonaltruist (N) is paired with another nonaltruist and an altruist, respectively. Note that in these probabilities, the focal individual comes after the vertical line (called a "pipe" by font fundis) ($\Pr(\text{other}|\text{focal})$), unlike in the payoff expressions ($V(\text{focal}|\text{other})$).

In Box 3.1 we show that by combining the above conditional probabilities with the payoffs to altruistic behavior, the condition for altruism to increase in frequency is

$$\{\Pr(A|A) - \Pr(A|N)\}\, b > c. \tag{3.1}$$

Altruism can evolve when altruists are more likely to interact with altruists than defectors are to interact with altruists. If $\Pr(A|A) = \Pr(A|N)$, then the left side of Expression 3.1 is always zero, and altruism is not favored by natural selection, regardless of the size of the benefit b. However, if some

---

Box 3.1 Derivation of statistical altruism model
The fitness of an altruist is

$$W(A) = w_0 + \Pr(A|A)V(A|A) + \Pr(N|A)V(A|N).$$

This would apply to any model with two strategies. Let's substitute in our altruism payoffs

$$W(A) = w_0 + \Pr(A|A)(b - c) + \Pr(N|A)(-c)$$
$$= w_0 + \Pr(A|A)b - c\{\Pr(A|A) + \Pr(N|A)\}.$$

Because $\Pr(A|A) + \Pr(N|A) = 1$,

$$W(A) = w_0 + \Pr(A|A)b - c.$$

Likewise, the fitness of a defector becomes

$$W(N) = w_0 + \Pr(A|N)b + \Pr(N|N)(0)$$
$$= w_0 + \Pr(A|N)b.$$

Cooperators will increase in frequency when

$$W(A) > W(N)$$
$$\Pr(A|A)b - c > \Pr(A|N)b,$$

which simplifies to Expression 3.1 in the main text.

---

process makes individuals more likely to interact with individuals like themselves, then altruism has a fighting chance.

So where do these probabilities of interaction, $\Pr(A|A)$ and $\Pr(A|N)$, come from? To answer this question, we need to add some more biology to the model. Evolutionary ecologists typically identify three kinds of mechanisms through which positive assortment of altruists might occur.

1. Limited dispersal. In many species, offspring disperse a limited distance from their natal territories. This means that like types will interact more often than chance, without any need for recognition or conditional social interaction.

However, limited dispersal is a two-edged sword—if dispersal is very limited, then the global population structure is not what is relevant to selection. We take up this issue in more detail at the end of this chapter.

2. Kin recognition. If animals recognize and preferentially interact with kin, then altruists will tend to interact with other altruists. This is the main topic of this chapter.

3. Behavioral bookkeeping. If interactions can continue for more than a short time, then a record of past behavior may allow altruists to discriminate other altruists from non-altruists. This is the main topic of Chapter 4.

Interesting effects may also arise when two or more of these mechanisms coexist. We will eventually touch upon a couple of these interactions.

3.3   Common descent and inclusive fitness

In 1963 and 1964, W. D. Hamilton published three papers that revolutionized evolutionary biology.[17] Hamilton's insight, previously expressed less formally by J. B. S. Haldane,[18] was that kinship could lead to the evolution of altruism, since an allele which codes for altruism could selectively help other copies of itself if altruistic behavior was preferentially directed towards kin. This is sometimes called *inclusive fitness*, the effects of an individual's actions on every individual's numbers of offspring, weighted by relatedness.[19] Mechanically, we calculate inclusive fitness as the sum of an individual's own fitness (personal fitness) and the fitness of its relatives, weighted by the coefficients of relatedness between each relative and the focal individual.

Let's apply the mechanism of common descent to the statistical altruism model above. This will convert the abstract probabilities $\Pr(A|A)$ and $\Pr(A|N)$ into more biologically meaningful expressions. Assume that individuals interact with kin in a prisoner's dilemma. There is a chance r that the individuals possess the same allele (strategy) by common descent. This is the coefficient of relatedness r for the category of kin which are interacting. There is an additional chance, however, that individuals share alleles (strategies) because of the frequency of the strategy in the population as a whole. Thus the chance that an altruist interacts with another altruist is

$$\Pr(A|A) = r(1) + (1 - r)p,$$

where p is the frequency of the altruist allele in the population as a whole. The conditional probabilities for the other types of interactions are derived similarly:

$$\Pr(N|A) = r(0) + (1 - r)(1 - p)$$
$$\Pr(N|N) = r(1) + (1 - r)(1 - p)$$
$$\Pr(A|N) = r(0) + (1 - r)p.$$

Notice that $\Pr(A|A)+\Pr(N|A)=1$ and $\Pr(N|N)+\Pr(A|N)=1$—the conditional probabilities for each strategy must sum to one. We call this the *haploid kin-selection* model because kinship influences probabilities of interaction, and each individual phenotype is specified by a single allele. Note that when $r = 0$, these expressions yield random interaction, just as in Chapter 2. When $r = 1$, types instead assort perfectly.

Substituting these expressions into Expression 3.1 gives

$$\left(\underbrace{\{r + (1 - r)p\}}_{\Pr(A|A)} - \underbrace{\{(1 - r)p\}}_{\Pr(A|N)} \right) b > c,$$

which quickly simpifies to

$$rb > c.$$

This expression is known as *Hamilton's rule*. In this expression, r is the coefficient of relatedness, the chance the actor and recipient share an allele by common descent. This is one for identical twins, one-half for full-siblings, one-quarter for half-siblings, one-eighth for cousins, and so on. b and c have the meanings they had in the additive prisoner's dilemma. The rule says: When the benefit, discounted by the relatedness of the actor and recipient, is greater than the cost to the actor, altruism can evolve.

In our experience, students usually understand Hamilton's rule as a universal law that applies to any organism on any planet. But as theoretical biologists are well aware, the rule doesn't actually apply exactly to *any* organism, at least on this planet. We begin this chapter by presenting two puzzles that we hope will shake your confidence that you understand Hamilton's rule. We then derive it in a rigorous way which reveals the assumptions on which it is based, and when we are finished you will see that Hamilton's rule *is* widely useful for understanding the evolution of social behavior. We go through this exercise for two reasons. First, Hamilton's rule definitely doesn't apply in some situations, and it's important to know when. Second, and perhaps more important, understanding why it sometimes works and why it sometimes fails will sharpen your understanding of how natural selection shapes behavior.[20]

3.3.1   Tearing it down

What exactly is wrong with Hamilton's rule? If you think hard about it, there are important aspects that do not make much sense. Here are two puzzles that highlight the problems.

Washburn's fallacy.   What exactly is r? Some people understand r as the proportion of genes in common over the entire genome. Of course, this number can't be right. If the allele is common, individuals may share it even if they aren't sisters (this is easy to see from our definition of the haploid kin-selection model). However, this misunderstanding of r is common and leads to an infamous fallacy. Sherwood

Washburn, a famous biological anthropologist, argued in the late 1970's that Hamilton's rule must be wrong because most alleles are very common.[21] Genetic data indicate that the vast majority of loci in the human genome are invariant. That is, at these loci, we all have the same genes. Thus, Washburn argued, Hamilton's rule says that we should all be universal altruists and only slightly more altruistic with close kin. And it gets worse. Genetic data also show that humans and chimpanzees share most of their genes. We no doubt show more differences with dogs, and yet more with roses. Does Hamilton's rule say we should be more altruistic with chimps than we are towards dogs and roses?

Most students think they know what's wrong with this argument. First, the only locus that matters is the locus at which the altruism allele resides. Second, r is the chance two individuals share the allele by *common descent*. Those selfish genes only care about real copies of themselves, not counterfeits! Unfortunately, this doesn't help us much. Common descent begins with the mutation that started the line of descent for a particular gene. Selection or drift then increases or decreases the number of copies of this original gene, while mutation introduces new non–identical-by-descent versions of the gene. Population genetics theory tells us that at equilibrium the number of nonidentical copies of alleles is approximately equal to:

$$4N\mu + 1.$$

N is the size of the population, and μ is the mutation rate.[22] Measured mutation rates in big mammals like humans are typically around 10^{-6}. Given this mutation rate, the number of unique non–identical-by-descent alleles for different population sizes is given as follows:

Population size (N)	Number of alleles
25000 (2.5×10^4)	1.1
2.5×10^5	2
2.5×10^6	11

Unless mutations are very common or populations are very large, it turns out that most genes *are* identical by descent! Was Washburn right after all?

Suicide and the single altruist. J. B. S. Haldane was once asked if he would risk death to save a drowning brother. "No," he quipped, "but I would to save two brothers, or eight cousins." Brian Charlesworth[23] imagined another paradox of Hamilton's rule which echoes this story. He asked us to consider a species of bird in which young have the option of staying behind and helping their parents care for the next season's young rather than going out and trying to found their own nests. In this particular species, a situation arises each generation allowing an individual to sacrifice its own life to save the lives of four of its younger full siblings. Thus $b = 4$, $c = 1$, and $r = 0.5$. According to Hamilton's rule, this behavior should evolve ($4 \times 0.5 > 1$), yet it cannot. In any individual in which the mutation arises, the allele will be destroyed.

3.4   Rediscovering Hamilton's rule

Clearing up these conundrums requires deriving Hamilton's rule in a way that exposes the assumptions on which it is based. We'll begin to reconstruct Hamilton's rule, using more-detailed genetic models of the evolution of altruism. At each step, we'll assume only what we need to in order to make progress. At the end, we'll have Hamilton's rule in its familiar form. We'll see why it gives rise to the two puzzles and why and when we can use it to understand the evolution of social behavior.

3.4.1   Covariance genetics

We will derive Hamilton's rule using, as Alan Grafen put it, a "pleasing way of doing population genetics"[24] introduced by George Price. The drawback of Price's method is that it tends to be more "pleasing" to mathematicians than biologists. But its beauty is that it applies to any system of inheritance. Being very general, it must be modified to make it useful for any particular biological (or cultural) problem. Nevertheless, Price's method provides a number of useful intuitions about the evolution of social behavior. We will use it to derive a very general form of Hamilton's rule. Then we will see how several assumptions allow us to simplify the general expression until we have Hamilton's rule in its usual form. Again.

Instead of deriving the Price equation in a rigorous fashion, we'll just provide a heuristic justification for why it is true. This will save time and reduce the chance that we'll lose any of you along the way. But don't worry—we'll do it the hard way later in the book (in Chapter 6), when it will be more useful.

Recall the original viability selection recursion we derived in Chapter 1:

$$p' = p\frac{V(A)}{\bar{w}},$$

where p is the frequency of allele A in the population, $V(A)$ is the average fitness of allele A, and $\bar{w}$ is the average, fitness for the population as a whole. Let's subtract p from both sides, making it a difference equation, and then multiply both sides by the mean fitness:

$$\bar{w}\Delta p = pV(A) - p\bar{w}. \tag{3.2}$$

The right side of Equation 3.2 looks like something you may have seen before. The covariance (cov) between two random variables x and y is

$$\text{cov}(x, y) = \text{E}(xy) - \text{E}(x)\,\text{E}(y).$$

Here E denotes an expectation, or long-run average value, of a random variable. The covariance of two variables is just the expectation of their product minus the product of their expectations.

It is helpful now to define some new terminology. First, we're going to give each the individual in the population a unique number i (like a national identification number) in the set $1, 2, 3, \ldots, n$. Define

w_i fitness of individual i
p_i frequency of allele A in individual i
n number of individuals in the population

For our haploid model, an individual i with allele A has $p_i = 1$, while an individual with any other allele has $p_i = 0$. For more complicated genetic systems, p_i could take on other values. In a diploid organism, it could be 0, 1/2, or 1. In a haplodiploid organism, p_i could take on different values for males (0, 1) and females (0, 1/2, 1). These definitions imply new definitions for some old friends. Since $\bar{w}$ is just the average individual fitness, and p is just the average individual allele frequency, we can now redefine $\bar{w}$ and p in terms of w_i, p_i and n:

$$\bar{w} = \mathrm{E}(w_i) = \frac{1}{n} \sum_i w_i$$

$$p = \mathrm{E}(p_i) = \frac{1}{n} \sum_i p_i.$$

Substituting these expressions into our difference equation (Expression 3.2) gives

$$\bar{w}\Delta p = \left\{ \frac{1}{n} \sum_i p_i \right\} V(A) - \left\{ \frac{1}{n} \sum_i p_i \right\} \left\{ \frac{1}{n} \sum_i w_i \right\}.$$

Note also (we're getting close!) that

$$V(A) = \frac{\sum_i p_i w_i}{\sum_i p_i}.$$

Recall that $V(A)$ is the mean fitness of the A allele. To compute this, we first sum the fitnesses of all individuals, weighted by the number of A alleles they have (this is the numerator). This sum is the sum fitness of A alleles. To get the average, we then divide this sum by the number of A alleles in the population, which is just the sum of the individual frequencies of A alleles in individuals (this is the denominator). Substituting this expression for $V(A)$ into our growing difference equation gives

$$\bar{w}\Delta p = \left\{ \frac{1}{n} \sum_i p_i \right\} \left\{ \frac{\sum_i p_i w_i}{\sum_i p_i} \right\} - \left\{ \frac{1}{n} \sum_i p_i \right\} \left\{ \frac{1}{n} \sum_i w_i \right\}.$$

This simplifies to yield

$$\bar{w}\Delta p = \underbrace{\left\{ \frac{1}{n} \sum_i p_i w_i \right\}}_{\mathrm{E}(p_i w_i)} - \underbrace{\left\{ \frac{1}{n} \sum_i p_i \right\}}_{\mathrm{E}(p_i)} \underbrace{\left\{ \frac{1}{n} \sum_i w_i \right\}}_{\mathrm{E}(w_i)}.$$

Now you can see that the right side is *exactly* the covariance between individual allele frequency (p_i) and individual fitness (w_i):

$$\bar{w}\Delta p = \mathrm{E}(p_i w_i) - \mathrm{E}(p_i)\,\mathrm{E}(w_i) = \mathrm{cov}(w_i, p_i).$$

This is the *Price equation*, and it yields an extremely useful result. While we just proved that the usual replicator dynamic we use for genetic selection is the same as the covariance version, it turns out that this is just one case of the Price equation. Any system of differential reproduction will be a special case of this expression. You may also know that a covariance is equal to a variance times the regression coefficient:

$$\mathrm{cov}(x, y) = \mathrm{var}(x)\frac{\mathrm{cov}(x, y)}{\mathrm{var}(x)} = \mathrm{var}(x)\beta(y, x),$$

where $\beta(y, x)$ is the slope of the regression of y on x (the slope of the best-fit line predicting y given x). This means we can represent the response to selection for any system with the variance in allele frequencies and the regression of fitness on individual genotype:

$$\bar{w}\Delta p = \mathrm{var}(p_i)\beta(w_i, p_i).$$

Note that we have made no assumptions about linear fitness functions, and yet all that matters at any one point in time is the linear regression coefficient and the variance. Although the derivation above assumed a haploid model, it turns out that the Price equation applies to any ploidy (e.g. diploid, haplodiploid). And for the empirically minded, $\mathrm{var}(p_i)$ and $\beta(w_i, p_i)$ might be things we have some hope of estimating in the real world.

3.4.2   Step 1: Additive fitness effects

We will make good use of the Price equation. First, we will use it to derive a more general form of Hamilton's rule. What we are doing is making one assumption at a time, seeing what kind of "rule" it yields, and then making an additional assumption, until we arrive at Hamilton's rule. This way, we will see what assumptions are required to get us to $rb > c$, as well as how each assumption affects the more general conditions for the evolution of altruism.

Let's begin with the general form of the Price equation and add one assumption, additive fitness effects. Additive means that being helped three times is three times as good as being helped once (Figure 3.1). This is probably rarely true in the real world. Most ecologists and economists think that most resources have diminishing returns—being helped three times is less than three times as good as being helped once. Nevertheless, it's hard to get anywhere in the algebra without this assumption. We'll reexamine it later, and see why it may not be such a bad assumption after all.

Now we have to define a function for individual fitness, w_i. Assuming additive effects of helping and being helped, we have

$$w_i = w_0 + y_i b - h_i c. \qquad (3.3)$$

This is just the linear PD all over again (Table 3.1). This equation says that the fitness of individual i is given by the baseline fitness (w_0), plus the probability that individual i

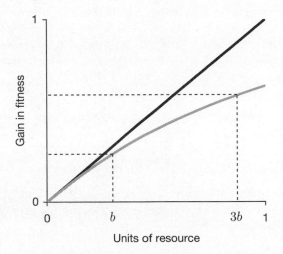

Figure 3.1: Additive fitness effects. The black line shows purely additive fitness returns on resources. Each additional unit of resource on the x-axis adds the same amount of fitness on the y-axis. The gray line shows a resource with diminishing returns. Each additional unit of resource adds a smaller amount of fitness than the previous unit, such that the gain in fitness after receiving $3b$ units of resource is much less than three times the gain after receiving b units.

receives aid (y_i) multiplied by the value of aid (b), minus the probability of giving aid (h_i) multiplied by the cost of giving aid (c).

In Box 3.2, we show that this definition of fitness leads to the following condition for the altruism allele to increase in frequency:

$$b\frac{\text{cov}(y_i, p_i)}{\text{cov}(h_i, p_i)} > c. \tag{3.4}$$

This looks a lot like Hamilton's rule, but with the ratio of the covariances substituted for r. To see what Expression 3.4 tells us, imagine that the covariance between the probability of giving aid and having the altruism gene, $\text{cov}(h_i, p_i)$, is a positive constant. (This just means that the more altruism

---

Box 3.2 Derivation of first condition for altruism to increase

Substituting the fitness expression, Expression 3.3, into the Price equation gives

$$\bar{w}\Delta p = \text{cov}(w_0 + y_i b - h_i c, p_i).$$

The covariance can be decomposed across addition, just like a sum, so we can now write

$$\bar{w}\Delta p = \text{cov}(w_0, p_i) + b \times \text{cov}(y_i, p_i) - c \times \text{cov}(h_i, p_i).$$

Since the covariance between individual genotype and baseline fitness (the first term on the right) is zero (constants do not vary, so they also cannot covary),

$$\bar{w}\Delta p = b \times \text{cov}(y_i, p_i) - c \times \text{cov}(h_i, p_i).$$

The altruism allele will increase in frequency when $\Delta p > 0$, which implies

$$b \times \text{cov}(y_i, p_i) - c \times \text{cov}(h_i, p_i) > 0.$$

After a little rearranging, the above yields the Expression 3.4 in the main text.

---

genes one has, the higher is h_i). Then, as the covariance between the probability that an individual receives help and the probability it carries the altruistic gene increases, the ratio of covariance increases. Thus Expression 3.4 says the higher the covariance between having an altruism gene and receiving aid, the more likely it is that altruism can evolve. This makes sense. If $\text{cov}(y_i, p_i) = \text{cov}(h_i, p_i)$, then altruists receive as much aid as they give, and all that matters is the magnitude of b relative to c. As $\text{cov}(y_i, p_i)$ decreases, the ratio approaches zero, and so this is a fractional multiplier something like the r in Hamilton's rule. If help is doled out at random, $\text{cov}(y_i, p_i)$ will equal zero, the ratio will equal zero, and altruism cannot possibly evolve. If having the altruism allele is a good predictor of being helped, then the ratio will be greater than zero, and if it is big enough, altruism can evolve. If we are willing to call the ratio of covariances "r," then we have Hamilton's rule already. But note that the ratio says nothing about relatedness. We need two more assumptions to get us there.

3.4.3 Step 2: Additive genetics

In order to make further progress, we need to assume that the genetic system doesn't matter. That is, we need to assume that phenotype (how the organism behaves, h_i) is a linear function of genotype (p_i). This implies

$$h_i = a + kp_i, \tag{3.5}$$

where a and k are constants which define the exact linear relationship. The parameter a defines the baseline amount of helping, when an individual has no altruism genes. The slope k determines the impact of each additional altruism gene on the probability of helping. When $a = 0$ and $k = 1$, there is an exact one-to-one mapping between the presence of altruism alleles and the frequency of helping. For other values, the relationship is less strict. Suppose, for example, we have a diploid organism. Then p_i can take on the values 0, 1/2, and 1. Defining h_i as we did above means that an individual with $p_i = 1/2$ is half as likely as one with $p_i = 1$ to help. Figure 3.2 shows this relationship graphically.

What about the covariance of being helped and genotype? We can write a function for this by assuming that whoever is helping our focal individual i obeys the same linear function, but conditioned on the helper's own genotype, p_j. Therefore,

$$y_i = a + kp_j.$$

Now we have functions that relate the genotype of the actor and the recipient to probabilities of helping and being helped. But what have we excluded with the assumption of additive genetics? Sickle cell, with it's heterozygote advantage? That's out. Any locus with dominance is out as well. In fact, we've probably excluded most of the loci in a diploid organism.

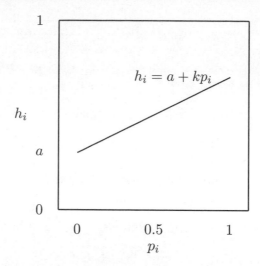

Figure 3.2: The probability of helping, as an additive function of individual genotype. Assuming no dominance or other nonlinear effect of gene frequency, this function is a straight line $h_i = a + kp_i$.

The assumption allows us to make progress through the algebra, though. With the assumption of additive genetics we show in Box 3.3 that the altruism allele will now increase when

$$b\beta(p_j, p_i) > c. \tag{3.6}$$

Now we have yet another version of Hamilton's rule, and at least this one has something to do with relatedness, albeit in a clumsy way. Expression 3.6 says that it is the slope of the regression line of p_j (the probability that the recipient has the altruistic allele) on p_i (the probability that the donor has it) that impacts the evolution of the altruistic allele. Those not familiar with regression may find this a little opaque. The regression line is the line that gives the best prediction (expected value) of the dependent/outcome variable (here, p_j) for each value of the independent/predictor variable (here, p_i).

This is hard to say, but easy to visualize. Figure 3.3 shows four regression lines of p_j on p_i for four imaginary

---

Box 3.3 Derivation of second condition for altruism to increase

Using Equation 3.5, we can expand the covariance between individual genotype and the probability of helping:

$$\text{cov}(h_i, p_i) = \text{cov}(a + kp_i, p_i) = k \times \text{cov}(p_i, p_i).$$

The variance of a random variable x is defined as $\text{E}(x^2) - \text{E}(x)^2$. A little inspection proves this to be the same as $\text{cov}(x, x)$. Thus the covariance of a variable with itself is its variance, and it follows that

$$\text{cov}(h_i, p_i) = k \times \text{var}(p_i).$$

Similarly, for the recipient,

$$\text{cov}(y_i, p_i) = \text{cov}(a + kp_j, p_i) = k \times \text{cov}(p_j, p_i).$$

We need to put these back together to see how far our assumption of no nonlinear gene effects has gotten us. Recall the general version of Hamilton's rule we derived before:

$$b \frac{\text{cov}(y_i, p_i)}{\text{cov}(h_i, p_i)} > c.$$

With our work just now, this becomes

$$b \frac{k \times \text{cov}(p_j, p_i)}{k \times \text{var}(p_i)} > c.$$

The constant k divides out, and the covariance over the variance is a regression coefficient, yielding the condition in the main text (Expression 3.6).

---

diploid populations. Imagine we sample a bunch of pairs of interacting individuals and assay their genotypes. We then calculate the average (expected) genotype of the recipient for each genotype of the donor. The line through these points is the regression line of p_j on p_i. The different lines in Figure 3.3 correspond to different strengths of relationship under two different overall frequencies of the altruism allele. When $\beta = 0$ (plots a and c), knowing p_i tells us nothing about p_j,

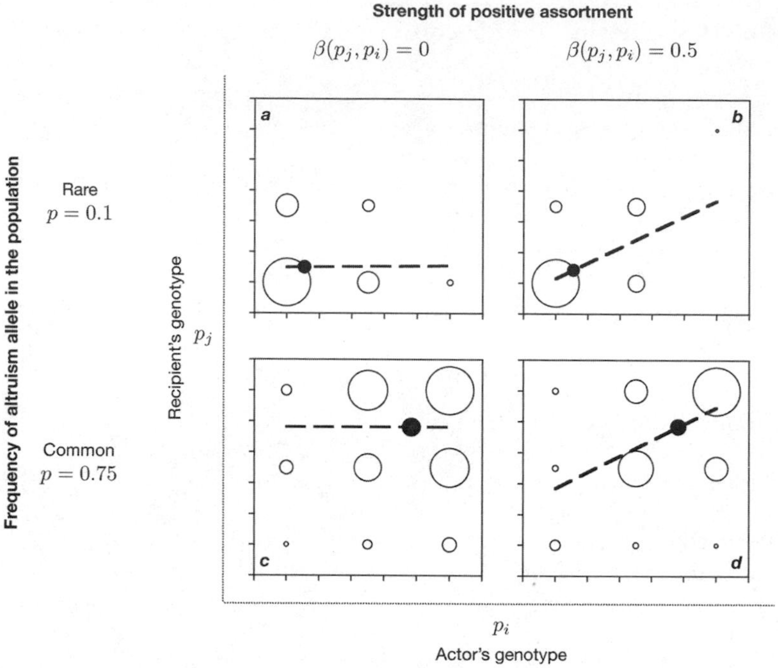

Figure 3.3: How positive assortment relates to the regression of recipient's genotype (p_j) on actor's genotype (p_i). The size of the white circles represents the number of pairs at each combination of p_i and p_j. The black circles in each plot represent the mean genotype of actors and recipients, which is p in each case. The dashed lines are the regressions of p_j on p_i. The slopes of these lines are the beta coefficients important in Expression 3.6.

and so the regression line is horizontal, passing through p, the population mean. Why? Because when there is no assortment based on genotype, the best guess of whom individual i will interact with is just the population mean, p.

It is possible to further specify this coefficient $\beta(p_j, p_i)$. The equation for a regression line is

$$E(y|x) = E(y) + \beta(y, x)\{x - E(x)\}$$

where this is the regression of y on x.[25] Any of the lines in

Figure 3.3 would be

$$\mathrm{E}(p_j|p_i) = \mathrm{E}(p_j) + \beta(p_j, p_i)\{p_i - \mathrm{E}(p_i)\}$$
$$= p + \beta(p_j, p_i)\{p_i - p\}.$$

By solving for the beta coefficient, we find that the regression coefficient in Expression 3.6 must be

$$\beta(p_j, p_i) = \frac{\mathrm{E}(p_j|p_i) - p}{p_i - p}.$$

You can see that the above expression contains deviations from the average frequency of the altruism allele in the population, p. If help were doled out at random, $\mathrm{E}(p_j|p_i) = \mathrm{E}(p_j) = p$, and so altruism cannot evolve because the value of β would be zero. If the recipient is no more likely to have the altruism allele than is a random member of the population, the regression coefficient is zero. It is the presence of positive deviations from p that allows altruism to evolve. So at least this version of Hamilton's rule (Expression 3.6) says that we want to know the likelihood that the recipient has the altruism allele, given that the donor has it. But we still don't have anything about common descent. Lots of other situations could generate a high β. What we need now is a way to calculate β from genealogies. When does $\beta(p_j, p_i) = r$?

3.4.4 Step 3: No selection for altruism

If you just read the title of this section and did a double-take, good. The final assumption we need to reach Hamilton's rule is that there is no selection at the locus for the altruism allele. Since we are interested in the evolution by natural selection of genes for altruism, this should be troubling. But trust us, it will be okay in the end.

Define r as the fraction of genes identical by descent (IBD) in both individuals i and j. This is the usual definition. With an outbred population, we have all been taught to compute r with a genealogy by counting the steps between the individuals and multiplying by one-half for each step (in the case of humans and other diploid organisms). This is because in a diploid organism each allele has an even chance of being inherited at each step. We repeat this process for each common ancestor and add the resulting products together to get r.

But this will only work if there is no selection favoring either allele. Otherwise, the chance that an allele for altruism is passed on will differ from that for other alleles at the same locus. It is easy to understand this in the most extreme case in which offspring who inherit the altruism allele do not survive to reproduce (as in Charlesworth's paradox of the missing altruists). In this case, even though meiosis may be fair and shunt altruism alleles into gametes exactly half the time, differential selection will make altruism less common in descendants than alternative alleles. Given no selection, however, it is easy to calculate the probabilities of interest, and so this is what we usually assume. From the form of Hamilton's rule we most recently derived (Expression 3.6), we need to know $E(p_j|p_i)$. Let $p_{j,\text{IBD}}$ equal the proportion of genes identical by descent (IBD) that are altruism genes. Let $p_{j,\overline{\text{IBD}}}$ equal the proportion of genes not IBD that are altruism genes. If interactions are between kin with a coefficient of relatedness of r, then

$$E(p_j|p_i) = rp_{j,\text{IBD}} + (1 - r)p_{j,\overline{\text{IBD}}}.$$

This expression arises from the fact that r is defined as the fraction of genes IBD. Given that this is true, the proportion of individual j's genes that are altruism genes must be the weighted average of those IBD with individual i (a proportion r of all genes, altruism or not) and those not IBD (a proportion $1 - r$ of all genes). As always, keep in mind that j might be an altruist, even if i and j are completely unrelated.

Let's figure out how to compute $p_{j,\text{IBD}}$. We need to know the proportion genes IBD in j that are altruism genes. If j and i have $r = 0.5$, then some proportion of these shared genes may be altruism genes. We want the expected proportion, given unbiased Mendelian inheritance. Let's assume a diploid organism for this example. If $p_i = 0$, then obviously none of the altruism alleles in j can be IBD because i has no altruism genes to be identical with. In this case, $p_{j,\text{IBD}} = 0$. If $p_i = 0.5$ (heterozygote individual), then we have no way of knowing which of these alleles is IBD. Since both alleles have a fair chance of being inherited at each stage in the genealogy, the chance that any given gene that is IBD is an altruism gene must be $1/2$. Thus $p_{j,\text{IBD}} = 0.5$ in this case. Finally, when $p_i = 1$, then any gene IBD in j must be an altruism gene, since i has no other alleles. Thus $p_{j,\text{IBD}} = 1$ in this case. You can see that, assuming all the things we've had to assume to get this far, the proportion of genes IBD that are altruism genes, given that the frequency of altruism genes in individual i is p_i, is simply p_i:

$$p_{j,\text{IBD}} = p_i.$$

The final thing we need to know is the proportion of genes not IBD that are altruism alleles. This is simply the population frequency of the altruism allele, p:

$$p_{j,\overline{\text{IBD}}} = p.$$

Thus, we can rewrite $\text{E}(p_j|p_i)$ as

$$\text{E}(p_j|p_i) = rp_i + (1 - r)p. \tag{3.7}$$

Solving for r gives

$$r = \frac{\text{E}(p_j|p_i) - p}{p_i - p}, \tag{3.8}$$

which is equal to the $\beta(p_j, p_i)$ that we derived earlier. We now have a way to compute $\beta(p_j, p_i)$ from genealogies, and so we finally have Hamilton's rule:

$$br > c.$$

Note that this also means that r is a special case of a regression coefficient. When (1) fitness effects are additive, (2)

phenotype is a linear function of genotype, and (3) there is
no selection, r is the slope of the best-fit line predicting the
genotype of the recipient (p_j), given the genotype of the donor
(p_i).

3.4.5   What is r?

We've come far enough now that we can spell out more clearly
what the r in Hamilton's rule is and is not.

r is not $\Pr(A|A)$.   It is also easy to see from Equations 3.7
and 3.8 that r is most definitely *not* the chance that the recipient has a given allele, given that the donor has it. This common mis-definition of r arises from a basic misunderstanding.
The recipient and donor can be alike simply because the allele is common in the population (p is large). Instead, r is
the *increased* similarly of the donor and recipient *relative* to
the population mean. Only in extreme cases, as when $p \approx 0$
and $p_i = 1$, will $r = \mathrm{E}(p_j|p_i)$.

**r is the extent to which actor's genotype predicts re-
cipients' genotypes.**   We showed a fairly standard derivation of r as a regression coefficient, as Hamilton presented it.
The slopes of the lines in Figure 3.3 are measures of r. In
calculus-speak, r is the partial derivative of p_j with respect to
p_i, $\partial p_j / \partial p_i$: the rate of change in recipients' genotypes as a
function of actors' genotypes. One important consequence of
this interpretation is that *negative* relatedness is possible. If
being an altruist predicts that the recipient is less likely than
chance to have the altruism allele, the slope of the best-fit line
will be negative. You might think this is a weird and unlikely
circumstance, but it might be important in some cases.[26]

**r is a statistic, not a number specific to any pair of
individuals.**   Relatedness in Hamilton's rule is a measure

of how actors' genotypes predict recipients' genotype, across many individuals. It is a statistic, not a data point.

r is not identical to Wright's r. Under the assumptions above, the r in Hamilton's rule will take the same value as the coefficient of relationship defined by Sewall Wright.[27] But this is a special case. In general, what matters to selection in Hamilton's rule is not conceptually the same as "relatedness," and this is why the "r" relevant to the evolution of altruism can in fact be negative (see just above) or be zero, even when social groups are comprised entirely of clones (see end of this chapter).

3.5 Justifying Hamilton's rule

We've just seen that in order to get Hamilton's rule, we had to assume additive fitness effects, no dominance (or other gene interactions), and no selection. Table 3.3 summarizes these steps and the form of Hamilton's rule arising from each.

Given these assumptions, which are always violated in real organisms, can Hamilton's rule ever be used? We think the answer is definitely "yes." Let's go through each of the assumptions and see why we can still use Hamilton's rule in its usual from.

3.5.1 Additive fitness effects and additive genetics

Most of the time, biologists believe that selection on individual alleles is weak. Complex traits are affected by alleles at many loci, and this means that the effect on fitness of an allele at any given locus is probably quite small. A convenient feature of small effects is that we can often just use a linear approximation of them. We have been using a similar approximation all along to determine stable equilibria. If changes in the frequency of a strategy are small enough, then the slope

Table 3.3: Steps needed to derive Hamilton's rule and the condition for the evolution of altruism at each.

Step	Assumption	Form of Hamilton's rule
0	Price equation	$\mathrm{cov}(w_i, p_i) > 0$
1	Additive fitness effects	$b\dfrac{\mathrm{cov}(y_i, p_i)}{\mathrm{cov}(h_i, p_i)} > c$
2	Additive genetics	$b\beta(p_j, p_i) > c$
3	No selection	$br > c$

of the line tangent to the actual (nonlinear) dynamics is all we need to know. The same principle applies here, for the same reason. If fitness effects are weak, then a linear approximation will be an excellent estimate of the actual functional form. The same argument saves us from worrying about gene interactions like dominance. Unless such effects have huge effects on fitness, the effect of substituting any one allele for another will be small enough to use a linear estimate.

However, when the behavior in question has life-or-death consequences, or it is impossible to characterize it with an additive model, something more general than Hamilton's rule will be necessary. In these cases, it is often possible to use the haploid kin-selection framework we derived earlier, a generalized equation like Equation 3.7, or a subtle approach like Taylor and Frank's[28] application of the chain rule to personal fitness functions (Box 3.4).

3.5.2 No selection

We don't really mean no selection at all, but weak enough selection that the bias for the altruism allele (or the non-altruism allele) is quite small from one generation to the next. When this is true, then calculating r for close relatives will still work as a very good approximation of what we really want to know: the increased chance that the recipient has the allele given that the donor has it. This is why Washburn was wrong: Hamilton's rule is an approximation for *close* relatives. For distant relatives, even weak selection will matter, since it will have had long-term effects on the frequencies of alleles in the population. As we have seen, it is also the solution to Charlesworth's paradox of the missing altruists.

3.5.3 No inbreeding

While we didn't mention inbreeding much above, it can also mess up how we compute r. Fortunately, we can usually ignore inbreeding. In many reasonably sized animal populations, matings between close relatives are rare, and we are usually interested in using Hamilton's rule for close relatives. In some species though, inbreeding is the norm rather than the exception. In these cases, there is nothing to do but figure what selection will do given the pattern of inbreeding.[29] Under some patterns of inbreeding, luckily, $rb > c$ is still a robust expression of the evolutionary dynamics.[30]

3.6 Using Hamilton's rule

After all this grinding, it would be useful to have examples of how to use Hamilton's rule to build models of the evolution of social behavior. First, we introduce a family of models concerning cooperative breeding. These models bear a deep similarity to usual models of parent-offspring conflict and parental investment. They are direct applications of Hamilton's rule. Second, we touch upon the important topic of the scale of

Box 3.4 Optimization approach to inclusive fitness

Taylor and Frank (1996) demonstrated a method of treating inclusive fitness when asking what genotype maximizes individual fitness. This approach is sometimes called "direct fitness" or "neighbor modulated fitness." It is subtle and not easily grasped by most people. Nevertheless, it is worth knowing about. Here we show how to derive Hamilton's rule via this method. Assume that individual fitness is a simple function of genotype:

$$w_i = w_0 + p_j b - p_i c,$$

where p_i is the genotype (probability of giving aid) of individual i. The key conceptual insight is that this function implies an additional function to relate p_j to p_i. Thus when natural selection attempts to maximize individual fitness, how it does so will depend upon this latent function. In order to find the value of p_i that maximizes individual fitness, we find the derivative of w_i with respect to p_i and find where it equals zero or, as in this case, whether it is always increasing or decreasing. The derivative of individual fitness with respect to individual genotype is (via the chain rule)

$$\frac{\mathrm{d}w_i}{\mathrm{d}p_i} = \underbrace{\frac{\partial w_i}{\partial p_i}}_{\mathbf{A}} + \underbrace{\frac{\partial w_i}{\partial p_j}}_{\mathbf{B}} \underbrace{\frac{\mathrm{d}p_j}{\mathrm{d}p_i}}_{\mathbf{C}}.$$

The chain rule decomposes a total rate of change into component parts. Term **A** is the direct marginal effect of individual genotype on individual fitness. Here, this is $-c$, the marginal cost of altruism. **B** is the marginal fitness effect of partner genotype on individual i's fitness. This is the marginal benefit of altruism, b. Finally, **C** is the rate of change in p_j with respect to p_i, which is formally like the slope p_j on p_i, the regression definition of relatedness we derived in the main text. Thus

$$\frac{\mathrm{d}w_i}{\mathrm{d}p_i} = \underbrace{\frac{\partial w_i}{\partial p_i}}_{-c} + \underbrace{\frac{\partial w_i}{\partial p_j}}_{b} \underbrace{\frac{\mathrm{d}p_j}{\mathrm{d}p_i}}_{r} = rb - c,$$

under the same assumptions that allowed us to derive Hamilton's rule. This expression is positive provided $rb > c$, and then the genotype that maximizes fitness is $p_i = 1$.

competition and how it alters the circumstances under which Hamilton's rule applies in its usual form.

3.6.1 Reproductive skew

Sometimes, groups of individuals breed cooperatively, several individuals collaborating in producing and caring for offspring. These individuals then share the total reproductive output of the nest, burrow, or colony. In eusocial animals like honeybees, the distribution of the shares is very uneven, with one or a few females monopolizing reproduction. In other animals, like wild turkeys, and a few human societies, brothers may share a single female mate. Biologists sometimes refer to the degree of equality or inequality in access to the cooperative breeding opportunities as *reproductive skew*. In this section, we'll explore the evolution of reproductive skew as an applied example of Hamilton's rule.

Imagine a species in which pairs of individuals can cooperate in breeding. Let's suppose two males who share a single female mate. Dominant individuals select the best nest sites first. A lone individual at a good site will have an expected number of offspring of 1. This is our baseline. Subordinates can take a low-quality nest where their expected number of offspring will be $1 - s$. However, the subordinate also has the option of joining the dominant individual at a good nest, and if it does, the combined productivity of the nest will lead to an average of $1 + j$ offspring between them. Assume the dominant individual regulates the allocation of mating opportunities and therefore controls the expected share of this $1 + j$ each male enjoys. Let the proportion of reproduction allocated to the subordinate individual be p. This proportion allocated to the subordinate is often termed the *concession*.

These assumptions allow us to write payoffs to each individual, nesting jointly or singly:

	Nesting singly	Nesting jointly
Dominant	$V_D(S) = 1$	$V_D(J) = (1-p)(1+j)$
Subordinate	$V_S(S) = 1 - s$	$V_S(J) = p(1+j)$

Finally, assume that the pair is related symmetrically by a coefficient of relatedness r.

Let's ask when it is in the subordinate's interest to join a dominant's nest. The cost to the subordinate of joining is the difference between nesting singly and nesting jointly. Similiarly, the benefit to the dominant will be the difference between nesting jointly and nesting singly. We compose the incremental cost and benefit above into Hamilton's rule to find out when natural selection favors joining by the subordinate:

$$r\Big(V_D(J) - V_D(S)\Big) > \Big(V_S(S) - V_S(J)\Big)$$

$$r\Big((1-p)(1+j) - 1\Big) > \Big(1 - s - p(1+j)\Big). \qquad (3.9)$$

$$\underbrace{}_{\text{incremental benefit to D}} \qquad \underbrace{}_{\text{incremental cost to S}}$$

Written another way, this becomes

$$V_S(J) + rV_D(J) > V_S(S) + rV_D(S).$$

The left side is the inclusive fitness of the subordinate when he joins the nest. The right side is his inclusive fitness when he nests alone.

Simplifying Expression 3.9, we see that it is in the subordinate's interest to join when the share of reproduction p satisfies the condition

$$p > \frac{1 - s - rj}{(1 + j)(1 - r)}. \qquad (3.10)$$

If the subordinate does not get at least this share of reproduction, it would be better for him to go it alone.

We can use Expression 3.10 to map out the range of ecological circumstances that may lead to joint nesting. First, consider the conditions under which the subordinate will want to join, regardless of the concession he receives. This will be true when the right side of Expression 3.10 is less than zero, which is true whenever

$$1 - s < rj.$$

This condition says that if the payoff to a subordinate nesting alone is sufficiently low, it can pay to join the dominant purely for the inclusive fitness benefits (rj). Keep in mind here that a subordinate who receives no concession always receives an inclusive fitness payoff of $r(1)$. It is the increment rj that is extra. Imagine, for example, cases in which yearlings are very unlikely to find and defend their own nest (s is large in this case). Even if they get no share of reproduction that year $(p = 0)$, it may be better to help their parents or older siblings reproduce than to try their own luck because the inclusive fitness benefits (rj) exceed their own expected reproduction $(1 - s)$ if they nest on their own.

Second, consider when always to nest alone, regardless of the concession p. This will be the case when

$$(1 - s) + r(1) > 1 + j.$$

The left side is the inclusive fitness to a subordinate who nests singly, and the right side is the inclusive fitness of a subordinate who nests jointly and monopolizes reproduction $(p = 1)$. When $j + s < r$, it may be better for related individuals to reproduce separately because the benefits of cooperative breeding are modest compared to the productivity of singleton nests. This may occur when good nesting sites are plentiful. Unless there are substantial cooperative breeding benefits (j large), it makes sense for each member of the family to found his or her own nest.

Next, consider when it might pay for the subordinate to join but only if it gains some concession ($1-s > rj$ and $1-s+r < 1 + j$). Notice that if $r = 0$, Expression 3.10 is satisfied only when the subordinate's share of reproduction exceeds the ratio of the payoff to nesting singly $(1 - s)$ to the payoff to nesting jointly $(1 + j)$. Thus if the synergies generated by nesting jointly are large enough, natural selection may favor joint nesting without any kin selection at all, and the share of reproduction necessary to make profitable for subordinates declines as these synergies increase.

What about the dominant's point of view? The subordinate may want to nest jointly, but it may not be in the dominant's interest to allow it. The dominant individual will want to nest jointly as long as the concession it has to provide to the subordinate satisfies

$$ r\Big(V_S(J) - V_S(S)\Big) > V_D(S) - V_D(J) $$

$$ p < \frac{j - r(1 - s)}{(1 + j)(1 - r)}. $$

This is the biggest concession the dominant is willing to make. This is greater than the amount the subordinate requires when

$$ \frac{1 - s - rj}{(1 + j)(1 - r)} < \frac{j - r(1 - s)}{(1 + j)(1 - r)}, $$

which simplifies to become

$$ j > 1 - s. \tag{3.11} $$

If this condition is not satisfied, the dominant will never provide a big enough concession sufficient to motivate the subordinate to join. Doing so would make the joint nest unprofitable to the dominant. Otherwise, the dominant will concede the smallest amount necessary to urge the subordinate to join, and this will be defined by (from Expression 3.10)

$$ p^\star = \frac{1 - s - rj}{(1 + j)(1 - r)}. $$

When the concession equals $p^\star$, the subordinate is indifferent to joining or nesting alone. Thus the dominant will cede a tiny fraction more than this amount.

Assuming Expression 3.11 is satisfied and it makes sense for the dominant to persuade the subordinate to join, what

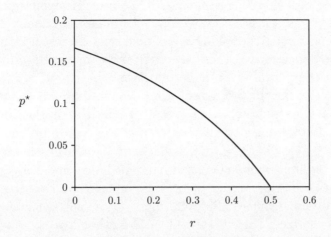

Figure 3.4: The minimum fraction of reproduction necessary to persuade the subordinate to join, $p^\star$, as a function of relatedness, r. In this example, $s = 0.75$ and $j = 0.5$.

effect does relatedness have on the subordinate's share of reproduction, $p^\star$? Figure 3.4 plots $p^\star$ as a function of r when $j > 1 - s$. Greater relatedness reduces the concession needed to make it worthwhile for the subordinate to join the nest. As r increases, the gains to the dominant are more important to the subordinate's inclusive fitness. Eventually (at $r = 0.5$, in this example), r is great enough that the subordinate is willing to join, even when he receives no share of the nest's reproduction. If his solitary breeding options are unattractive enough, and he can enhance his relative's reproduction sufficiently, inclusive fitness justifies foregoing any attempt at personal reproduction and aiding the dominant. This is an extreme outcome, however.

Thus this model predicts that (1) we will observe cooperative breeding only when both individuals benefit, (2) the dominant will concede no more than it has to to keep the subordinate from leaving, and (3) relatedness increases reproductive skew: as r increases, p decreases.

This very basic skew model demonstrates how breeding options to different individuals and relatedness may affect cooperative breeding decisions. This model is sometimes referred to as "classic skew theory." But there are many other ways to model the biology. To take one interesting variant of the above model, there are often more than two individuals in the potential pool of cooperative breeders. A market may arise in which different dominant individuals "bid" for a subordinate to join their nests. When this is the case, it turns out that genetic relatedness has no effect on the outcome.[31]

This sudden change in the strategic landscape is a general feature of bargaining models: subtle changes in the structure of the model, like adding one more social partner, can dramatically alter the results. There are other models in which the skew results from explicit struggle for control within the nest. These models can make quite different predictions about the distribution of reproduction in cooperatively breeding species. Parent-offspring and sibling competition models sometimes have the same kind of sensitivity to structural assumptions.[32] Thus, which models are relevant depends upon carefully matching the assumptions of the model to the natural history of the species of interest. The general and robust Hamilton's rule does not necessarily imply general and robust predictions about behavior.

3.6.2   Local competition

Positive assortment among altruists is necessary (but not sufficient) for the invasion and maintenance of altruism. Most of this chapter has been about common descent, kinship, as a mechanism for creating positive assortment among the phenotypes of social partners. One way for social groups comprised of kin to form is through kin recognition—you look or smell like me.[33] A more basic mechanism is limited dispersal, called "viscosity" by Hamilton. If offspring do not move very far from their birthplace, many available social partners will be relatives. However, it turns out there is an important limitation to this mechanism.

To see how limited dispersal can be incorporated into the models we have already examined, we need to back up and reconsider the structure of the population. In all previous models we implicitly assumed that all population regulation is *global* (some population geneticists call this *hard selection*[34]). That is, each local group (or pair) competes globally with every other group for spots in the next generation. Processes that limit population size depend on the size (or density) of the whole population. It is hard to imagine that population regulation is literally global. Competition over food and mates is usually a local process. However, regulation may be approximately global if after social interaction but before population regulation a large fraction of individuals disperse to other local groups. As a consequence, social groups thoroughly mix before local density dependence reduces local populations (Figure 3.5(a)). At the global scale, this is nearly equivalent to a model in which all local groups are dissolved each generation, offspring inhabit a global "clubhouse," and new local groups are formed by randomly sampling offspring. This is what we conveniently assumed in every model up to this point. Reproduction is curtailed for the entire clubhouse, meaning each offspring is equally likely to suffer mortality due to crowding, regardless of which local group their parents came from.

You might think this model is rather unrealistic. It is. However, when there is substantial migration among local groups, this panmictic model approximates more realistic population structure quite well. It predicts, for example, that the proportion of genetic variation among groups will be approximately $1/n$, where n is the size of local groups (we discuss this in more detail in Chapter 6). Measured estimates of the variation among groups are often quite close to this for large animals like ourselves.

Now suppose that each local group send only a number $n - m$ of its members to the clubhouse. These individuals reproduce, and all their offspring undergo uniform regulation. The m individuals who stay in local groups reproduce, and

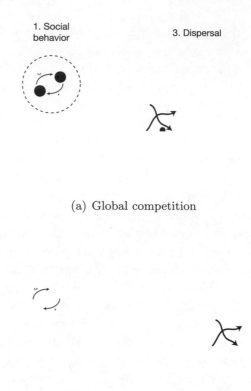

(a) Global competition

(b) Local competition

Figure 3.5: Filled circles represent altruists. Hollow circles represent nonaltruists. Numbers next to each dashed circle are counts of individuals in each local group at each stage in the life cycle. We assume in these examples that the environment limits the number of adults in each local group to two. Dispersal transfers half of each local group. (a) When dispersal occurs after social behavior but before population regulation, altruists can spread. This is because groups with more altruists contribute more migrants to the population migrant pool. New local groups are drawn largely at random from migrants, and once population regulation acts to reduce the number of offspring each generation, altruists are likely to compete with individuals other than those they originally cooperated with. (b) When dispersal occurs after population regulation (pictured) or is limited (discussed in the text), altruism is harder to evolve. Altruists might help one another, but they must later compete with one another before dispersal mixes the survivors of local groups.

their offspring must compete locally for m spots in the next generation. The remaining $n - m$ spots in each local group will be filled with offspring sampled from the clubhouse. The ratio m/n is a measure of the extent of *local population regulation*, what population geneticists sometimes call *soft selection*. As m approaches n, local groups become more and more isolated from one another with respect to natural selection. If $m = n$, then the average fitness in any other local group is irrelevant because your offspring will only compete with offspring of the other members of your own local group.

What does this have to do with limited dispersal or viscosity? There are different ways local competition can occur. Soft selection will arise if population regulation occurs after social interaction but before dispersal (Figure 3.5(b)).[35] When this is the case, a fixed number will disperse, and social behavior has no effect on the number of emigrants. Local competition can also result from a patchy environment and limited migration among patches or a continuous environment and limited dispersal (a "viscous" environment). When offspring do not disperse very far, the scale of population regulation becomes more local. If your kids will move next door to you, then they will have to compete only with others who also move next door, not the entire population, many of whom will reside in distant habitats. Thus an individual's neighbors are also his or her most direct competitors, provided dispersal is limited. If they are also his or her kin, they may be genetically and phenotypically similar. But even if all brothers and sisters behave altruistically towards one another, the whole can contribute only n offspring to the next generation.

It turns out that this severely limits the circumstances under which altruism can evolve. Altruists need to find one another for social interaction but avoid competing with one another afterwards. When population regulation is global, altruists are competing against the global average fitness, as is clear from the viability recursion in Chapter 1, $p' = pW(\mathrm{A})/\bar{w}$. When population regulation is local, the average fitness in the population becomes less relevant than the average fitness in the local group.[36] Consider the difference equation for the frequency of an altruism allele, p, when all local groups indexed i are regulated locally[37]:

$$\Delta p = \mathrm{E}\left(p_i(1 - p_i)\frac{w_i(A) - w_i(N)}{\bar{w}_i}\right).$$

This just means that it is the average increase or decrease across all groups that will give us the population frequency in the next generation. Now notice that the only local groups in which there will be a change in the local frequency of the allele, p_i, are those groups in which there is variation. If the groups consist of pairs, for example, then the only groups that matter have one altruist and one nonaltruist. In other groups there is no variation for selection to act on. This gives us

$$\Delta p = \underbrace{\Pr(A, A)(0)}_{p_i=1} + \underbrace{\Pr(A, N)\frac{1}{2}\left(1 - \frac{1}{2}\right)\frac{-c - b}{(b - c)/2}}_{p_i=0.5}$$

$$+ \underbrace{\Pr(N, N)(0)}_{p_i=0}$$

$$= -\Pr(A, N)\left(\frac{1}{2}\right)\frac{b + c}{b - c}.$$

This quantity is always less than zero, so the altruism allele always decreases in frequency. Regardless of how strongly altruists are associated with other altruists ($\Pr(A, N)$), only mixed groups affect the frequency of altruism alleles, and in mixed groups altruists do worse than nonaltruists. This is the most extreme case, but it illustrates the serious issue local competition creates for altruism.

W. D. Hamilton was the first person to understand how viscosity limits the evolution of altruism.[38] Others later derived generalizations of Hamilton's rule for cases of local competition.[39] This work shows that local competition acts to reduce positive assortment, as we define it in this chapter. To see this, let ℓ be the extent to which population regulation

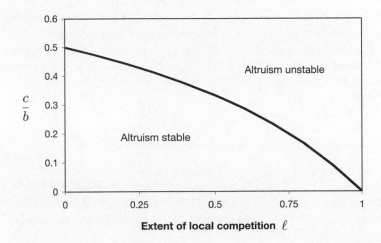

Figure 3.6: Graph of the critical cost-benefit ratio for altruism to be an ESS as a function of the extent of local competition ℓ (Expression 3.12). In this example, $r = \bar{r} = 0.5$. When populations are regulated locally, altruism is much harder to evolve. Local regulation can result from limited migration or the timing of dispersal. See Figure 3.5.

is local. When $\ell = 1$, within-group competition is everything. When $\ell = 0$, we have the kind of models we have seen up to this point. If local groups comprise individuals with average relatedness $\bar{r}$, and altruists help siblings with a coefficient of relatedness r, altruism increases in frequency provided (Box 3.5):

$$\frac{r - \ell\bar{r}}{1 - \ell\bar{r}}b > c. \tag{3.12}$$

When $\ell = 0$, we recover Hamilton's rule. When $\ell = 1$, positive assortment is greatly reduced if there is general similarity (relatedness, $\bar{r}$) built up within local groups (Figure 3.6). Note that if the groups are pairs, then $\bar{r} = r$, and local competition will quickly reduce any selective advantage of altruism.

What is happening here? As ℓ increases, the relevant average genotype by which positive assortment must be judged

Box 3.5 Viscosity and positive assortment

Suppose local groups comprise individuals with average coefficient of relatedness $\bar{r}$. Pairs of kin with relatedness r within these groups can behave altruistically. Let ℓ be the extent to which the population is regulated locally rather than globally. The regression definition of Hamilton's rule says that positive assortment is measured by

$$\beta(p_j, p_i) = \frac{\mathrm{E}(p_j|p_i) - \mathrm{E}(p_j)}{p_i - \mathrm{E}(p_i)}.$$

When population regulation is global, $\mathrm{E}(p_j) = \mathrm{E}(p_i) = p$, the global frequency of the altruism allele. However, when regulation is local, $\mathrm{E}(p_j)$ and $\mathrm{E}(p_i)$ may depend upon i's genotype, because the actor may be related to all individuals in the local group. The regression coefficient becomes

$$\beta(p_j, p_i) = \frac{\mathrm{E}(p_j|p_i) - \left\{(1-\ell)p + \ell\,\mathrm{E}(p_k|p_i)\right\}}{p_i - \left\{(1-\ell)p + \ell\,\mathrm{E}(p_k|p_i)\right\}},$$

where $\mathrm{E}(p_k|p_i)$ is the expected genotype of the average individual in individual i's group. If $\bar{r} > 0$, $\mathrm{E}(p_k|p_i) = (1-\bar{r})p + \bar{r}p_i$. Substituting this for $\mathrm{E}(p_k|p_i)$, using the definition of $\mathrm{E}(p_j|p_i)$, Expression 3.7, and simplifying yields

$$\beta(p_j, p_i) = \frac{r - \ell\bar{r}}{1 - \ell\bar{r}}.$$

Substituting the new value of β into Expression 3.6, we obtain Expression 3.12 in the main text.

approaches the mean in the local group. Consider the extreme case when social groups comprise of clones ($r = \bar{r} = 1$). If competition is entirely local, then there can't be any positive assortment within the group because every individual in the group has the same genotype. Selection will not favor helping in a prisoner's dilemma when this is the case, even though social partners are close relatives.

Guide to the Literature

Prisoner's dilemma. The original concept is old, going back at least to Thomas Hobbes' 1660 book, *The Leviathan.* The term itself is attributed to Albert Tucker, who invented the prison framing for a seminar given at Stanford University. **Hamilton's rule.** Hamilton's 1964 derivations of his rule recognized r as a regression coefficient. Confusion and debate about what the rule and r mean have a long history. See Dawkins 1979. Grafen's 1985 geometric interpretation of r is clarifying and illustrates the relationship between inclusive fitness and selection in general. More recently, Frank's 1998 book on social evolution favored an approach to the evolution of altruism that avoids direct use of Hamilton's rule. See Box 3.4 for an example of this approach. Empirical studies like Clutton-Brock's (see Clutton-Brock 2002 for citations) and Gadagkar 2001 have contributed to feelings that the role of relatedness has been oversold. **Price equation.** Price's 1970 paper derived the basic covariance difference equation. His 1972 paper applied it to the evolution of altruism. Frank's 1995 review of the intellectual contributions of Price and his equation is worth reading. Taylor 1990 and Grafen 2005 expanded Price's formula to include classes of individuals. **Reproductive skew.** The first skew model is due to Vehrencamp in 1979. The literature has expanded since to include explicitly competitive tug-of-war models. See Johnstone 2000 and Reeve 1998 for accessible reviews. Skew models are deeply similar to models of sibling rivalry and parent-offspring conflict. For those, see Mock and Parker's 1998 book, as well as Robert Trivers' original 1974 essay. Hanna Kokko's 2003 analysis of the assumptions underlying skew models is highly recommended. **Local competition and viscosity.** The distinction between global (hard) selection and local (soft) selection has long been a concern of ecologists and population geneticists. Wallace 1968 is credited with the terminology. Hamilton 1964 mentioned that limited dispersal might favor altruism, but in 1971 (see also

Hamilton 1996) he argued, verbally, that very viscous populations would have limited altruism, because of local competition. Hamilton's 1970 model of spite implicitly assumed perfectly soft selection and found altruism could not evolve. Boyd 1982 derived a version of Hamilton's rule under flexible local competition and found it to limit altruism. Later, Taylor 1992 and Wilson et al. 1992 simultaneously published analyses reaching the same conclusion. Queller's 1994 note explained how viscosity's reduction of altruism is consistent with inclusive fitness theory, broadly defined. Axelrod et al. 2004 provided a formal demonstration of synergy between limited dispersal and kin-recognition mechanisms. Gardner and West 2004 reviewed the theory of altruism in viscous populations, with particular attention to the evolution of spiteful behavior.

Problems

1. Hawk-Dove among relatives. The logic of inclusive fitness often allows us to dispense with a lot of game-theoretic flourishes by jumping right to Hamilton's rule. However, one must be careful about when this is justified. For example, the general form of Hamilton's rule for inclusive fitness is

$$rV(B|A) + V(A|B) > rV(B|B) + V(B|B),$$

where $V(A|B)$ is the payoff to a player using strategy A against an opponent playing strategy B. (a) In the case of a prisoner's dilemma, as defined in this chapter, show that the above rule simplifies to Hamilton's rule if Cooperate is strategy A and Defect is strategy B, giving the same result as the haploid kin selection model we used to derive Hamilton's rule. (b) We might also ask when, in competition over scarce resources, an animal will allow another related animal to have the resource without a fight. Assume the usual Hawk-Dove game, with $v > c$, such that it always pays to fight, on average, but that pairs of individuals are kin with an average coefficient of relatedness r. Show that the general inclusive fitness rule presented above predicts that playing Dove can invade when

$$r > \frac{v - c}{v + c}.$$

(c) Using the haploid kin selection model presented in the chapter, show that the above result is incorrect. Instead, Dove can invade when

$$r > 1 - \frac{c}{v}.$$

(*Hint*: Assume the frequency of Doves is approximately zero, but note that $\Pr(D|D) > 0$ still.) (d) Provide a verbal explanation of why Hamilton's rule gives the wrong answer in the case of the Hawk-Dove game. What assumption(s) does it violate?

2. Covariance assortment. In the chapter, we presented a general non genetic form of Hamilton's rule expressed in terms of conditional probabilities of interaction:

$$\{\Pr(C|C) - \Pr(C|D)\}b > c.$$

The difference of conditional probabilities measures the strength of assortment. Show that the most general covariance form of Hamilton's rule,

$$\frac{\text{cov}(p_i, y_i)}{\text{cov}(p_i, h_i)}b > c,$$

reduces to the condition above above when y_i and h_i are expressed in terms of the conditional probabilities of interaction ($\Pr(C|C)$, etc.). Assume a haploid model and that an individual with the altruism allele always helps and an individual without it never helps.

3. Who do you help? Assume that an organism has a fixed amount of resource, R, to divide among two individuals with coefficients of relatedness r_1 and r_2, where $r_1 > r_2$.

Receiving an amount of resource x increases the fitness of that individual $x^{\frac{1}{2}}$. (a) Suppose that an individual gives x_1 to the r_1 individual. Write down an expression for the inclusive fitness associated with this definition in terms of x_1, r_1, r_2, and R. (b)You learned in your calculus class that you can find the maximum of a function by setting the derivative of the function equal to zero. Use this technique to show that the inclusive fitness maximizing value of x_1 is

$$x_1 = \frac{R}{1 + \left(\dfrac{r_2}{r_1}\right)^2}.$$

Show that this expression says that the selection will favor giving some to each individual, but that the more closely related individual gets more. (c) Now, suppose that the fitness effects of receiving a donation of x is x^2. Use the same method as in (b) to show

$$x_1 = \frac{R}{1 + \left(\dfrac{r_1}{r_2}\right)}.$$

Show that this equation says that $x_1 < \frac{1}{2}$—give less to the more closely related individual. What's wrong here? What do you think is the inclusive fitness maximizing allocation in this case? Can you prove it? (*Hint:* Show that the derivative of inclusive fitness is negative for x_1 less than this value, and then think about what this says about the value of x that maximizes inclusive fitness.)

4. Frequency dependent assortment. In many species a combination of limited dispersal and a tendency for similar organisms to seek out similar environments leads to a patchy distribution of genotypes in space. In other words, similar types are more likely to interact than chance alone would dictate. Consider a species which interacts in pairs. Individuals can either behave altruistically, A, or selfishly, S, with the following payoffs:

| | Player 2 | |
Player 1	A	S
A	$b - c$	$-c$
S	b	0

Assume $b - c > 0$. The probabilities that each of the three possible pairings occurs are, respectively,

$$\Pr(A, A) = \frac{p^2(1 + a)}{1 + ap},$$

$$\Pr(S, S) = \frac{(1 - p)^2(1 + a)}{1 + a(1 - p)},$$

and

$$\Pr(A, S) = p(1 - p)\left(\frac{1}{1 + ap} + \frac{1}{1 + a(1 - p)}\right),$$

where a is a positive constant. Increasing values of a mean more assortative interaction. (a) Show that selfish behavior is evolutionarily stable. (b) Show that altruistic behavior is not evolutionarily stable. (c) Show that altruistic behavior increases at intermediate frequencies of altruists if a is large enough. *Hint*: Any intermediate value of p will do, but setting $p = 0.5$ makes the calculation easier.)

5. The stag hunt. In his great book on the nature of society, *The Leviathan*, the seventeenth-century British philosopher Thomas Hobbes (who wrote just after the bloody English Civil War) conceived of life in the state of nature as a prisoner's dilemma game in which rational people could not cooperate. The result, Hobbes said, was a "warre of all against all" resulting in life which was "solitary, poor, nasty, brutish, and short."

Writing almost 100 years later in (at the time) peaceful France, Jean Jacques Rousseau painted a very different picture of the state of nature in his *Discourse on Inequality*. Rousseau asked us to consider a population in which pairs of individuals have two options: they can hunt for "a stag" or for "hare." Hunting hare is a solitary activity, and an individual who chooses to hunt hare gets a small payoff, which we will label h, no matter what the other individual does. Stag hunting, however, requires cooperation. If both players hunt for the stag, they usually succeed and each gets a large payoff, which we will label s. However, a single individual hunting stag always fails and gets a payoff of 0. The resulting game, now called "the stag hunt," leads to a much sunnier, intrinsically more cooperative view of society than Hobbes' prisoner's dilemma.

Interestingly much the same argument has gone on in biology recently. Some biologists feel that there has been too much emphasis on the difficult problem of the evolution of altruism and as a result a too pessimistic picture of the possibilities of social cooperation. Instead, they think much social cooperation results from mutualism, and thus we should concentrate more on games like the stag hunt.

(a) Assume that there are two heritable strategies, hunt hare, H, and hunt stag, S, and that pairs are formed at random. Using the description of the game given above, write down the game matrix. Assuming that $s > h > 0$, determine the evolutionarily stable equilibria of the game and the basin of attraction of each.

(b) Suppose that interacting pairs are relatives with coefficient of relatedness r. Use the haploid sexual model to determine the conditions under which S and H are ESSs. Again determine the basins of attraction of each ESS. Describe the qualitative behavior of the model as relatedness increases. What do you think this model predicts about the relationship between kinship and mutualistic behavior in nature?

Notes

[10]Sherman 1977.

[11]Wilson 1975, page 3.

[12]Pennisi 2005.

[13]Hamilton 1964.

[14]Trivers 1971.

[15]Dugatkin 1990 modeled the example of predator inspection in shoaling fish as a cooperative behavior with a Hawk-Dove payoff structure. In such a situation, it is typical that there will be a stable mix of cooperators and non-cooperators in the population.

[16]See Clutton-Brock 2002 for a discussion of other routes to "cooperation."

[17]Hamilton 1963, 1964.

[18]Haldane 1955.

[19]Grafen 1984.

[20]Our treatment here closely follows Grafen's seminal 1985 paper.

[21]Washburn 1978.

[22]Kimura and Crow 1964. This model assumes no selection, but under selection there will be even fewer alleles present because an allele favored by selection can sweep the population.

[23]Charlesworth 1978.

[24]Grafen 1985.

[25]You can prove this to yourself by considering the more-familiar equation for such a line: $E(y|x) = a + \beta x$, where a is the y-intercept (the value of y when $x = 0$). Solve for a:

$$a = E(y|x) - \beta x.$$

Now take the expectation of both sides:

$$E(a) = E\{E(y|x)\} - E(\beta x)$$
$$a = E(y) - \beta E(x).$$

Finally, substitute a back into the original equation:

$$E(y|x) = E(y) - \beta E(x) + \beta x$$
$$= E(y) + \beta(x - E(x)).$$

[26]Hamilton 1970.

[27]Wright 1922.

[28]Taylor and Frank 1996. See also Frank 1998.

[29]Michod 1979, 1980; Uyenoyama 1984.

[30]Roze and Rousset 2004.

[31]Johnstone 2000; Reeve 1998 provides accessible reviews of the reproductive skew game-theoretic literature.

[32]Mock and Parker 1998.

[33]Grafen 1990b.

[34]Wallace 1968.

[35]Taylor 1992.

[36]In Chapter 6, we will see that this effect can be understood as local competition reducing the between-group component of selection, which favors altruism.

[37]In Chapter 6, we show that this can be derived from the Price equation, but we let it stand as common sense for now.

[38]Hamilton 1971. See also Hamilton's preface to this paper in Hamilton 1996.

[39]Boyd 1982, Taylor 1992, Queller 1994.

Box 3.6 Symbols used in Chapter 3

Prisoner's dilemma

b Marginal benefit to recipient of altruistic behavior

c Marginal cost to provider of altruism

B Total marginal benefit to group resulting from action of an altruist

r Chance a given gene is identical by descent

Price equation

$\bar{w}$ Mean fitness in the population

$E(x)$ The expected value of a variable x

$cov(x, y)$ Covariance between the random variables x and y

n Number of individuals in the population

w_i Fitness of a given individual i

p_i Frequency of an allele in individual i

$var(x)$ The variance in a variable x

$\beta(y, x)$ The slope of the best-fit line predicting y, given x

Hamilton's rule

y_i Probability individual i receives aid

h_i Probability individual i provides aid

a A constant defining the amount of aid an individual provides when $p_i = 0$

k A constant defining the amount of additional aid an individual provides with a unit change in p_i

s Reduction in number of offspring a subordinate produces, relative to a dominant, when nesting alone

j Additional number of offspring joint dominant and subordinate produce together

ℓ Proportion of population regulation that is local as opposed to global

$\bar{r}$ Average pairwise relatedness in local groups

Chapter 4

Reciprocity

In most primate societies, individuals have long-lasting relationships with other members of their groups. Individual monkeys often have preferred grooming partners and reliable allies. If allies and grooming partners were always relatives, we could explain such cooperation with inclusive fitness and move on. But these cooperative relationships involve unrelated individuals. How can natural selection lead to such behavior in the absence of kinship?

One answer is reciprocity: I'll scratch your back if you scratch mine. For example, Robert Seyfarth and Dorothy Cheney[40] showed that vervet monkeys are more likely to respond to the recorded distress cries of monkeys who have recently groomed them than to the same animals when they have not helped them recently. In 1971, Robert Trivers introduced the idea to evolutionary biology that natural selection could favor altruism between unrelated individuals provided that animals (1) recognize one another individually, (2) can keep track of past interactions, and (3) can direct altruism to those who have helped them in the past.

The paper attracted some attention, but the idea really took off in 1981 when Robert Axelrod and W. D. Hamilton published a paper in *Science* that analyzed a formal model of reciprocal altruism. This paper, and Axelrod's 1984 book on the same topic, generated an avalanche of work on reciprocity that extends their work and calls into question some of their results. In this chapter we cover both the basic Axelrod-Hamilton model of reciprocity and some of the more important developments since, including the importance of mistakes, the effect of partner choice, indirect reciprocity, and the roles of reciprocity and altruistic punishment in solving collective action problems.

Along the way, this chapter also explains how to build and solve models of repeated interactions. Repeated interaction is important, since many social species have persistent group membership. Vervet monkeys see familiar kin and non-kin nearly every day. To adequately model the evolution of their social behavior, we need to include the effects of repeated interaction. For humans, of course, repeated interaction is crucial. You'll also learn how to build models of interactions among more than two individuals, so-called n-person games.

4.1   The Axelrod-Hamilton model

We explained in Chapter 3 that greater-than-chance association of altruists is necessary for altruism to evolve by natural selection. Common descent can create this positive assortment. The key question about the evolution of reciprocity is: can natural selection also design an altruistic strategy that uses previous behavior to generate positive assortment among altruists? To answer this question, Axelrod and Hamilton had to add two new features to the evolutionary games analyzed in the last chapter: (1) repeated interaction between the same individuals and (2) behavior that is contingent on the history of interaction. Here's how Axelrod and Hamilton modeled each feature.

Repeated interactions.   At first blush, you might think that a good approach would be to suppose that relationships last a fixed number of periods. However, in the words of

Richard Nixon, "that would be wrong." The reason is that assuming a fixed number of interactions leads to the "lame duck" effect. Lame duck legislators lose much of their power during their last term in office because they cannot make promises or threats about what they will do in the next legislative session. The same goes for evolution—mutants who do not cooperate in the last period will always be favored. For example, if all relationships last exactly three periods, then there is never any incentive to cooperate in the last turn, since good deeds performed in the last period cannot be rewarded and bad deeds cannot be punished. Once such mutants have spread and no one cooperates in the last turn, then mutants who do not cooperate in the next-to-last turn will also be favored. This corrosive logic continues until selection leads individuals to defect in every turn.

We don't observe the lame duck effect in most relationships because individuals aren't sure when the relationship is going to end. To incorporate this key aspect of repeated interactions, Axelrod and Hamilton assumed that relationships end with a constant probability each period. It is as if instead of a having fixed term limits, a special coin were flipped each election cycle. If the coin comes up heads, the legislator may run again. If it comes up tails, he may not. Thus the legislator (and his allies and enemies) never know when his or her term will really end. This is a very special model of repeated interaction. It implies that the fact that two individuals have been together for 10 years says nothing about the probably that they will still be together next year. This means that incentives to cooperate and defect do not change during the history of the interaction. When you solve Problem 5 at the end of this chapter, you will see that changing this assumption can have important effects.

Contingent Behavior. The essential feature of reciprocity is contingent cooperation—I cooperate if and only if you cooperate. To allow for such contingent behavior, we assume that strategies are *rules* that specify how to behave.

Strategies can be fixed, always specifying the same behavior. For example, the strategy ALLD always defects, and the strategy ALLC always cooperates. However, now strategies can also specify different behavior depending on what has previously occurred in the game. For example, a famous strategy called Tit-for-Tat (TFT) begins each series of interactions by cooperating. In each subsequent interaction in a series, TFT plays as its opponent played on the last interaction. We have already seen simple examples of such rules—Bourgeois plays Hawk if owner and Dove if intruder. In repeated interactions, rules are crucial.

4.1.1   Modeling repeated interactions

With these two assumptions, we proceed exactly as in the Hawk-Dove game. Imagine a very large population in which individuals use one of two strategies, ALLD and TFT. Each generation, a large number of pairs of individuals are sampled at random and play a prisoner's dilemma (PD) as defined in Chapter 3. Each interaction, there is a chance w that any given pair continues on to play the prisoner's dilemma and a probability $1 - w$ that their relationship terminates. Payoffs for all rounds are summed and become an individual's payoff for that generation. This game structure is called the *iterated prisoner's dilemma* (IPD).

The first task is to compute the expected payoffs for each pair of strategies. The payoff to a TFT individual interacting with another TFT on round 1 of their interaction is

$$V_1(TFT|TFT) = b - c.$$

In any round that follows, the payoff will be the same, since neither will voluntarily defect.

There is a chance w each round that another interaction occurs. Thus the expected payoff of each TFT individual during round 2 is $b - c$ multiplied by the chance that the interaction continues to round two, w; the expected payoff during round 3 is $b - c$ multiplied by the chance that the interaction continues three periods, w^2; and so on. Thus, the expected payoff of a TFT given that it interacts with another TFT is

$$V(TFT|TFT) = b - c + w(b - c) + w^2(b - c) + w^3(b - c) + \ldots$$

This is an infinite series in which the payoff in round n is multiplied by the probability of reaching that round, w^{n-1}. This looks awful, but as is shown in Box 4.1, it is easy to close the series and derive the following expression for the payoff of TFT versus TFT:

$$V(TFT|TFT) = \frac{b - c}{1 - w}. \tag{4.1}$$

The factor $1/(1 - w)$ is the average number of interactions between a pair of individuals, and $b - c$ is the payoff that each TFT player earns each period.

Next, let's calculate the expected payoff of TFT versus ALLD. On the first round, TFT cooperates and ALLD defects. This yields

$$V_1(TFT|ALLD) = -c.$$

On each subsequent interaction, TFT also defects, since ALLD defected on the previous round. So each additional round after the first yields a payoff of zero:

$$V(TFT|ALLD) = -c + w(0) + w^2(0) + w^3(0) + \ldots = -c.$$

The full payoff is just $-c$. Similarly, ALLD versus TFT receives

$$V(ALLD|TFT) = b + w(0) + w^2(0) + w^3(0) + \ldots = b.$$

Box 4.1 Closing an infinite geometric series
First, note that the series is of the form

$$x + wx + w^2x + w^3x + \ldots$$

Now factor out x, since it is in every term. This yields

$$x(1 + w + w^2 + w^3 + \ldots).$$

Thus all we need to close is the quantity in the parentheses (which is called a geometric series). Let D equal this infinite series:

$$D = 1 + w + w^2 + w^3 + \ldots \qquad (4.2)$$

Factoring gives

$$D = 1 + w(1 + w + w^2 + w^3 + \ldots).$$

Notice now that the quantity in parentheses is equal to D as we originally defined it (Equation 4.2). So we can substitute D for the parentheses:

$$D = 1 + wD.$$

Solving for D yields an expression for the average number of rounds interactions continue:

$$D = \frac{1}{1 - w}.$$

Remembering that $x = b - c$ and assembling the full series at top yields Equation 4.1 in the main text.

We now use these payoffs to determine whether TFT and ALLD are evolutionarily stable. A population in which TFT is common can resist rare ALLD invaders if

$$V(TFT|TFT) > V(ALLD|TFT).$$

Substituting in the payoffs we just calculated yields the following condition:

$$\frac{b - c}{1 - w} > b.$$

TFT almost always interacts with TFT and so gets $b - c$ for as long as the relationship persists. Rare defectors get b on the first turn and nothing thereafter. This condition can be rewritten as

$$wb > c,$$

which looks just like Hamilton's rule—just substitute w for r. This similarity makes sense because the parameter w expresses the stability of reciprocating pairs. Remember that cooperation can evolve only when cooperators receive more benefits of cooperation than defectors. Defectors enjoy the benefits of one cooperative act, at most, but pairs of TFT cooperate for as long as the interaction persists. Thus when pairs persist for a long time, most of the cooperative acts in a population occur within such pairs. This means that TFT individuals receive a disproportionate share of cooperation. This behavioral assortment sustains reciprocal altruism.

So, TFT is an ESS as long as relationships last long enough. What about ALLD? When ALLD is common, it can resist invasion by rare TFT individuals if $V(ALLD|ALLD) > V(TFT|ALLD)$. This requires that

$$c < 0.$$

Thus TFT cannot invade a population of ALLD—both TFT and ALLD are ESSs against the other. As you will show in Problem 1 (you are working the problems, aren't you?), these are the only stable equilibria. You will also see that there is a single unstable equilibrium that marks the boundary between the basins of attraction of ALLD and TFT.

So, our analysis tells us that populations can reach two quite different long run outcomes—no cooperation and full cooperation, depending on initial conditions. As we will see, in more complicated models, many other strategies can also be stable equilibria, and these strategies produce every pattern of behavior. This is not happy news. All we have to show for our work so far is the prediction that anything can happen—the kind of conclusion that gives theoreticians a bad name. A useful theory must tell us which equilibria are likely evolutionary outcomes and which are not.

4.1.2   Kinship and reciprocity

One thing that can make the reciprocal outcome more likely is kinship. Many people think of kin selection and reciprocity as alternative routes to the evolution of altruism. Among kin, kin-selection rules, and among non-kin, reciprocity drives altruism, when it can exist. This view is mistaken—there is a powerful synergy between kinship and reciprocity. In the last section we saw TFT cannot invade ALLD when rare. Here we will show that when reciprocity produces big long-term benefits, a little bit of kinship can go a long way toward helping reciprocal strategies invade noncooperative populations. This means that reciprocity is a much more likely evolutionary outcome than noncooperation. It also means we should expect *more* reciprocity among kin than among non-kin.

So far we have assumed that individuals are paired at random. TFT was just as likely to get paired with an ALLD as another TFT. If pairs are related, then this will not be true. We already saw the haploid kin-selection model in Chapter 3. In Box 4.2, we combine those probabilities of interaction with the iterated prisoner's dilemma model to show that TFT can now invade a population of ALLD when

$$r > \frac{1-w}{b/c - w}. \tag{4.3}$$

This expression is very cool. First, notice that if $w = 0$, we are right back to the pure kin-selection model, and if $r = 0$, we get back the pure-reciprocity result just derived. This means that we did the algebra correctly, which is good. But here's the cool part. Expression 4.3 says that there is a strong synergy between the two, so that a little bit of kinship allows TFT to invade a population of ALLD. All that is needed is enough nonrandom interaction to pair up rare reciprocators at the outset. Table 4.1 shows the value of r (calculated from Expression 4.3) that TFT will invade, for five values of w, with $b = 2$ and $c = 1$. When $w = 0$, we get the familiar kin-selection result—behaviors with a twofold benefit-cost ratio

---

Box 4.2 Condition for TFT to invade with kinship
Using the haploid kin-selection model (Chapter 3), we find that, respectively, the fitnesses of TFT and ALLD become

$$W(TFT) = \Pr(TFT|TFT)V(TFT|TFT)$$
$$+ \Pr(ALLD|TFT)V(TFT|ALLD) + w_0$$
$$= \{r + (1-r)p\}\frac{b-c}{1-w} - (1-r)(1-p)c + w_0$$

and

$$W(ALLD) = \Pr(TFT|ALLD)V(ALLD|TFT)$$
$$+ \Pr(ALLD|ALLD)V(ALLD|ALLD) + w_0$$
$$= (1-r)pb + \{r + (1-r)(1-p)\}(0) + w_0.$$

TFT will invade a population of ALLD when (recall that $p \approx 0$ here)

$$r\frac{b-c}{1-w} - (1-r)c > 0.$$

Solving for r yields Expression 4.3 in the main text.

Table 4.1: Threshold values of r for Tit-for-Tat to invade ALLD, for five values of w.

w	0	0.25	0.5	0.75	0.9
Required r	0.5	0.43	0.33	0.2	0.09

are only favored among full sibs. As w increases, however, less and less kinship is needed to destabilize ALLD. For $w = 0.9$ (only 10 interactions on average), r need only be about 1/10!

The synergy between kinship and reciprocity is a key part of the Axelrod-Hamilton model. In evolutionary game theory models, we typically assume that individuals are paired at random. This is not a bad assumption for many species, especially large, mobile mammals like humans, because there is ample gene flow among social groups. To a rough approximation individuals do interact at random when they interact with other members of their social group who do not share a recent common ancestor. However, there is often a small amount of background relatedness within groups, and a better approximation is to assume that there is a small tendency to interact with relatives. We see here that even small amounts of assortative interaction might allow reciprocal strategies to invade when rare and stabilize them when common. The reason is easy to see. When strategies interact at random and defection is common, there is no chance that individuals carrying rare reciprocating genes will meet. This means that the long-run benefits associated with sustained cooperation are irrelevant. Reciprocators get exploited, and that is that. However, when there is some assortative interaction, rare reciprocators do occasionally meet, and if the long-run benefits of cooperation are big enough, even a small amount of assortment can cause the average fitness of reciprocators to exceed the average fitness of defectors. This in turn suggests that even though both TFT and ALLD are ESSs, reciprocation is a much more likely evolutionary outcome than defection.

4.2 Mutants and mistakes

So far we have used TFT as the prototypical reciprocating strategy. Many other reciprocating strategies are possible, but in his 1984 book, Alexrod argued that TFT is the best reciprocal strategy. A key part of his argument was TFT's success in two computer "tournaments." Axelrod solicited a large number of strategies (instantiated as FORTRAN subroutines—*sic transit gloria!*) from various interested parties and then pitted these against each other in two round-robin tournaments. TFT won both tournaments, and so Axelrod concluded that TFT did well against a variety of different strategies. Since TFT could also increase when rare, and persist when common, it seemed like a good candidate for the best reciprocal strategy.

Subsequent research has taught us that TFT is not a good strategy in all environments. In fact, there is no single "best" strategy in the iterated prisoner's dilemma. The best strategy depends on what kinds of mutants are invading and what kinds of mistakes the common strategy makes. Different patterns of mutations and mistakes favor different strategies.

In this section, we first explain why the solution to the IPD depends on the kinds of rare invaders that are present. We then show how different kinds of mistakes favor different kinds of reciprocating strategies.

4.2.1   Is TFT an ESS?

Let's begin by adding a third strategy, ALLC, to the population. We know from Chapter 3 that ALLC cannot be an ESS—it can always be invaded by ALLD—so you might think that adding ALLC won't make any difference. But you'd be wrong.[41]   To see why, notice that since ALLC always cooperates and TFT always cooperates, $V(TFT|TFT) = V(TFT|ALLC) = V(TFT|TFT) = V(TFT|ALLC)$. This means that as long as ALLD is absent, ALLC and TFT have the same fitness. In the deterministic world of our model, there is no selection, and this means nothing happens. What this really means, however, is that we have left important stuff out of the model. Selection is not the only evolutionary process affecting gene frequencies, but normally we ignore processes like genetic drift[42] because they are too weak to be important.

But in this case there is no selection, and therefore drift will be important. Even if the population is initially dominated by TFT, the frequency of the two strategies will vary randomly. As long as TFT is common, ALLD will be excluded from the population. However, eventually drift will cause ALLC to exceed the threshold frequency $1 - \frac{c}{wb}$ (Box 4.3), and then ALLD will invade, leading to universal defection.

Box 4.3 The threshold frequency of ALLC that allows ALLD to invade

We can find the critical frequency of ALLC just as we found the critical frequency of Retaliator in Chapter 2. Let q be the frequency of ALLC. Then rare ALLD types will invade the TFT/ALLC mix when

$$V(TFT|TFT) < qV(ALLD|ALLC) + (1 - q)V(ALLD|TFT).$$

Remember ALLC and TFT have the same fitness when ALLD is rare. Then

$$\frac{b - c}{1 - w} < q\frac{b}{1 - w} + (1 - q)b$$

$$q > 1 - \frac{c}{wb}. \tag{4.4}$$

Figure 4.1 illustrates the dynamics of these three strategies. Notice that while ALLD always invades when $q > 1 - \frac{c}{wb}$, the system sometimes returns to another TFT/ALLC equilibrium. This is because, as long as there is at least a critical frequency of TFT (shown by points to the right of the dashed line in the figure), TFT increases each generation. This increase may be slow at first, but eventually it pulls the system back towards the line of TFT/ALLC equilibria. Thus the system may undergo cycles: periods of high levels of cooperation followed by the invasion of defectors. In the long run, the average frequency of ALLD might be quite high.

This simple example illustrates a key fact about repeated interactions. Namely, there are many different strategies that have the same fitness against each other. ALLC and TFT are members of a large class of "nice" strategies, so called because they are never the first to defect. Any nice strategy has the same payoff against any other nice strategy. They differ in how they deal with defection. If interactions go on long enough and a reciprocating strategy like TFT is common, defection can't compete. This means that any defections must be due to either rare mutants or cooperators making errors. Since the fitness differences between the nice strategies are

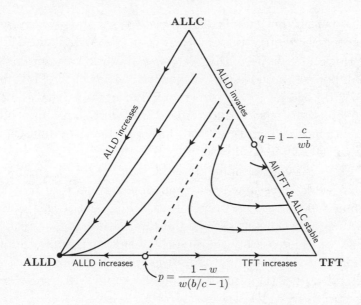

Figure 4.1: Ternary plot of the TFT, ALLC, ALLD system dynamics (for $b/c = 4$, $w = 0.5$). The dashed line represents is all points for which $W(TFT) = \bar{W}$. To the right of this line, TFT increases. To the left, it decreases.

solely due to defections, this means that the relative fitness of different nice strategies depends on the pattern of defections created by either mutation or mistake.

4.2.2 Any strategy can be destabilized

In fact, it is possible to show[43] that *every* strategy can be destabilized by the right combination of mutants, even though mutation is very weak. To see why, let's focus on two different reciprocating strategies: TFT and Tit-for-Two-Tats (TF2T), which cooperates on the first move but defects if its opponent has defected on the previous two turns. TF2T isn't obviously dumb—the late great John Maynard Smith submitted it in Axelrod's second tournament, and it finished in second place. TFT and TF2T have the same fitness playing each other, because neither defects without provocation, and so they get the cooperative payoff every turn, just like ALLC and TFT.

However, suppose that ongoing mutation maintains a small pool of rare invaders. When we consider how TFT and TF2T do against these rare types, it turns out that TF2T can sometimes invade and replace TFT but sometimes not, depending on which rare types are introduced by mutation. We will illustrate each case.

First consider a population in which TFT is common. So far we have assumed that mutations are so rare that invading types never encounter each other, and if this is the case, neither of these strategies invades (although, like ALLC, TF2T can increase by genetic drift). However, suppose mutation creates enough invaders that they occasionally meet each other. Let the frequencies of TF2T and ALLD be p and q, respectively. Then the fitness of TFT is

$$W(TFT) = w_0 + (1 - p - q)V(TFT|TFT)$$
$$+ pV(TFT|TF2T) + qV(TFT|ALLD).$$

The fitness of TF2T is

$$W(TF2T) = w_0 + (1 - p - q)V(TF2T|TFT)$$
$$+ pV(TF2T|TF2T) + qV(TF2T|ALLD).$$

The fitness of ALLD is

$$W(ALLD) = w_0 + (1 - p - q)V(ALLD|TFT)$$
$$+ pV(ALLD|TF2T) + qV(ALLD|ALLD).$$

TFT can resist invasion by TF2T if

$$W(TFT) - W(TF2T) > 0.$$

Since $V(TFT|TFT) = V(TFT|TF2T) = V(TF2T|TFT) = V(TF2T|TF2T)$, this condition becomes

$$V(TFT|ALLD) - V(TF2T|ALLD) > 0.$$

Table 4.2: Moves in a pair consisting of Tit-for-Tat (TFT) and Suspicious-Tit-for-Tat (STFT).

Strategy	Round 1	Round 2	Round 3	Round 4	
TFT	C	D	C	D	...
STFT	D	C	D	C	...

The relative fitnesses of TFT and TF2T depend only upon the relative difference in payoffs against ALLD. Other interactions are irrelevant. We already know $V(TFT|ALLD)$. The payoff to TF2T against ALLD is

$$V(TF2T|ALLD) = -c - wc + w^2(0) + w^3(0) + \ldots$$
$$= -c(1 + w).$$

Because TF2T cooperates a second time before it gives up, it pays, on average, an additional cost cw. Thus TFT is an ESS against TF2T whenever cooperation is costly. Since TFT is also an ESS against ALLD, TFT is an ESS when TF2T and ALLD invade.

Everything changes, however, if Suspicious-Tit-for-Tat (STFT) invades instead of ALLD. STFT is willing to cooperate but is initially suspicious. It begins by defecting but then plays just like TFT. The only interactions that matter, again, are those with the nasty rare strategy (STFT in this case). Thus, TFT will resist invasion by TF2T when

$$V(TFT|STFT) > V(TF2T|STFT).$$

The payoff to TFT against STFT is

$$V(TFT|STFT) = -c + wb - w^2c + w^3b - w^4c + \ldots. \quad (4.5)$$

This is because TFT begins by cooperating, but STFT begins by defecting. In the next round, each strategy copies its opponent's last move. This leads TFT to defect and STFT to cooperate. In the next round, they copy one another again. This trading of recriminations continues for the duration of the interaction (Table 4.2). Simplifying Equation 4.5 gives

$$V(TFT|STFT) = -c(1 + w^2 + w^4 + w^6 + \ldots)$$
$$+ b(w + w^3 + w^5 + w^7 + \ldots)$$
$$= (wb - c)(1 + w^2 + w^4 + w^6 + \ldots).$$

This infinite sum can be closed using the same trick introduced in Box 4.1, yielding

$$V(TFT|STFT) = \frac{wb - c}{1 - w^2}.$$

We calculate payoff to TF2T against STFT in the same way. On the first move, TF2T cooperates and STFT defects. On the next move, both cooperate. TF2T cooperates because it only defects after two defections. STFT cooperates because its opponent cooperated on the last move (Table 4.3). This leads to

$$V(TF2T|STFT) = -c + wb + w^2b + w^3b + \ldots$$
$$= -c + \frac{wb}{1 - w} = \frac{wb - c}{1 - w}.$$

Comparing TFT to TF2T, we see that TFT can be invaded when

$$V(TF2T|STFT) - V(TFT|STFT) > 0$$

$$\frac{wb - c}{1 - w} - \frac{wb - c}{1 - w^2} > 0,$$

which is true as long as $w > 0$. Therefore, TF2T can invade TFT, provided that STFT is maintained by mutation. TFT is never stable against TF2T, provided that STFT exists at low frequency. Because TF2T is less provokable, it can establish stable cooperation with STFT where TFT, with its hair trigger, cannot.

These examples illustrate that just knowing how each strategy does against each other strategy one at a time isn't

Table 4.3: Moves in a pair consisting of Tit-for-Two-Tats (TF2T) and Suspicious-Tit-for-Tat (STFT).

Strategy	Round 1	Round 2	Round 3	Round 4	
TF2T	C	C	C	C	...
STFT	D	C	C	C	...

enough. Every common strategy always has many potential competitors that have the same payoff against the common strategy as the common strategy has against itself. This means that the relative fitness of such strategies depends on how the common strategy and the invaders play against other invading strategies. Given certain restricted assumptions about mutation, some strategy may indeed be an ESS. But since biologists and social scientists rarely if ever know anything about what strategies are likely to invade, this ties our hands a bit. The lesson here is that it is very hard to say precisely what strategy is an ESS in any sufficiently iterated game, unless players make mistakes.

4.2.3 Mistakes determine the ESS

So far, we have assumed that individuals never make mistakes. They never defect when they intend to cooperate nor cooperate when they intend to defect. But suppose you plan to take a friend to the airport one morning, but you accidentally oversleep. You have defected, even though your intention was to cooperate. Such errors, even quite rare, can change the results we derived for the basic IPD, for exactly the same reason as multiple invading strategies did. In both the case of multiple invading strategies and in the case of mistakes, additional assumptions can lead some strategy or set of strategies to dominate in an otherwise indifferent world.

There are two kinds of errors to consider. We explore each in turn. We don't actually explain much of the mathematics in these sections because the calculations can be laborious and add little understanding of how the models work. Instead we explain the logic of errors and their effects.

Implementation errors. Imagine a pair of TFT players interacting in an IPD. Now suppose one of them makes an error, marked by the **X** in Table 4.4(a). We assume that player knows he made a mistake, which means that this is an *implementation* error, a mistake in carrying out an action, but not in knowing what has happened. This mistake will set off a chain of recriminating defections as the second player defects in response to the error, the first player then responds with defection to the second player's defection, and so on (as shown in Table 4.4(a)).

This sequence of mutual recrimination will return to cooperation only if one of the players erroneously cooperates when it intended to defect. Such erroneous cooperation is probably rare. Cooperation takes planning and coordination. In contrast, defection often entails simply doing nothing. How likely is it you accidently show up on the right day and accidentally take your friend to the airport? Its much easier to imagine a host of events which could prevent you from doing so. Thus it makes sense to assume that the reciprocal error, C instead of D, is much rarer than D instead of C.

This means that once sequences of recrimination begin, they will be unlikely to end before interactions stop. It is much more likely in fact that one of them will play D instead of C again and begin a sequence of mutual defection. This is bad news for TFT. If mistakes are common enough, they will significantly reduce $V(TFT|TFT)$, eventually allowing ALLD to invade.

One strategy people have proposed to deal with this problem is Contrite Tit-for-Tat (CTFT).[44] CTFT keeps track of *standing*. If a player cooperates, it gains good standing. If a player defects against a player in good standing, it earns bad standing. CTFT follows the rule: Cooperate with players in good standing; defect against players in bad standing; if I am

Table 4.4: Implementation errors and the evolution of reciprocity. The tables illustrate how implementation errors impact different reciprocal strategies. Each table shows sequences of behavior in a single pair. Implementation errors are marked by an **X** above the round number and a bold **D** in the behavior row. (a) Two TFT begin trading defections once one commits an implementation error. (b) Contrite-Tit-for-Tat (CTFT) is just like TFT when facing ALLD. (c) However, CTFT against itself is able to recover from implementation errors.

(a) Accidental defections hurt pairs of TFT.

				X			
	Round	n	$n+1$	$n+2$	$n+3$	$n+4$	...
TFT	Payoff	$b-c$	$b-c$	b	$-c$	b	
	Behavior	C	C	**D**	C	D	...
TFT	Behavior	C	C	C	D	C	...
	Payoff	$b-c$	$b-c$	$-c$	b	$-c$	

(b) Contrite-Tit-for-Tat resists exploitation by ALLD.

		n	$n+1$	$n+2$	$n+3$	$n+4$	...
	Round	n	$n+1$	$n+2$	$n+3$	$n+4$	...
CTFT	Payoff	$-c$	0	0	0	0	
	Standing	g	g	g	g	g	
	Behavior	C	D	D	D	D	...
ALLD	Behavior	D	D	D	D	D	...
	Standing	g	b	b	b	b	
	Payoff	b	0	0	0	0	

(c) Pairs of Contrite-Tit-for-Tat recover from implementation errors.

				X			
	Round	n	$n+1$	$n+2$	$n+3$	$n+4$	...
CTFT	Payoff	$b-c$	b	$-c$	$b-c$	$b-c$	
	Standing	g	g	b	g	g	
	Behavior	C	**D**	C	C	C	...
CTFT	Behavior	C	C	D	C	C	...
	Standing	g	g	g	g	g	
	Payoff	$b-c$	$-c$	b	$b-c$	$b-c$	

in bad standing, cooperate. All players are assumed to be in good standing when interactions begin.

Against ALLD, CTFT looks just like TFT. It begins by cooperating, but ALLD's defection against it puts ALLD into bad standing, leading to mutual defection until interactions end. In Table 4.4(b), we denote a player in good standing with a g and one in bad standing with a b. Standing changes after the behavior marked just before it is performed. CTFT can defect against ALLD without losing good standing. So CTFT does just as well against ALLD as does regular TFT (see Table 4.4(b)).

However, errors do not lead to endless recrimination when CTFT plays CTFT. Suppose, for example, that the first CTFT player in a pair commits an implementation error at the **X** in Table 4.4(c). This player is then in bad standing (b) immediately following, and so the other player can defect without falling into bad standing. The first player finds himself in bad standing, and so cooperates in order to escape it. Thus after one round of a "justified" one-sided defection, two CTFT players enjoy continuing mutual cooperation (Table 4.4(c)).

You can see the contrition in CTFT by noticing that the individual who made the mistake accepts a defection from his partner *without retaliating*. Suffering a defection this way while still cooperating also removes the payoff advantage the individual gained from making the mistake. If you sum up the payoffs in Table 4.4(c), you will see that the players end up trading defections and therefore have the same payoff. Since the partner cannot tell what the individual's intentions were—whether the defection was a mistake or a strike of opportunity—this costly acceptance of his partner's defections is a credible signal of good intent.

Perception errors. CTFT does well in the presence of implementation errors. However, it is vulnerable to another type of error. *Perception* errors can occur when an individual cooperates but his opponent thinks that he defected. In this kind of situation, CTFT will be in bad standing according to her opponent, but think she is in good standing. If CTFT doesn't know she made a mistake, she cannot be contrite and cooperate in the face of her opponent's defection. This will lead to the same problem TFT had with implementation errors. Table 4.5(a) provides an example, where the second CTFT player commits a perception error in round $n+1$. We have to keep track of each player's perceptions of himself and the other now, so each has two rows for his beliefs about standing. After the error, the first player then thinks he is in good standing, while the second thinks the first is in bad standing. This leads the second to commit what he believes is a justified defection, but the first sees it as an unjustified defection. This process repeats, leading to the same pattern of recriminations that TFT suffered.

One strategy that deals well with perception errors is Pavlov, so named by Martin Nowak and Karl Sigmund.[45] Pavlov cooperates on the first move, and is therefore a "nice" strategy like TFT and CTFT. Pavlov afterwards cooperates if both players cooperated in the previous round or if both players defected in the previous round. In all other cases, it defects. The idea here is that Pavlov changes its behavior when it receives a "bad" payoff. It defines mutual defection and being suckered (playing C to its opponent's D) as "bad."

Suppose the second of a pair of Pavlov players commits a perception error at the **X** in Table 4.5(b), thinking his opponent defected. The first player plays C again in the next round because, as far as he knows, both players cooperated in the previous round. But the second player plays D, since he thinks he was suckered in the previous round. The first player now gets the sucker's payoff, so he switches to defection in the following round, two rounds after the perception error. Now both players defect, and so three rounds after the error, they both switch back to cooperation. They have therefore recovered from the perception error.

Pavlov quickly recovers from perception errors. However, unlike CTFT, Pavlov is very vulnerable to nasty strategies

Table 4.5: Perception errors and the evolution of reciprocity. The tables illustrate how perception errors impact different reciprocal strategies. Each table shows sequences of behavior in a single pair. Perception errors are marked by an **X** above the round number and a bold **D** in the behavior row. (a) Contrite-Tit-for-Tat (CTFT) does poorly when perception errors are common because players end up with conflicting beliefs about standing. (b) Pavlov does better when perception errors are common. (c) Unfortunately, Pavlov is repeatedly exploited by simple strategies like ALLD.

(a) Pairs of Contrite-Tit-for-Tat fail to recover from perception errors.

	Round	n	**X** $n+1$	$n+2$	$n+3$	$n+4$	
	Payoff	$b-c$	$b-c$	$-c$	b	$-c$	
CTFT	Own standing	g	g	g	g	g	
	Other's standing	g	g	g	b	g	
	Behavior	C	**D**	C	D	C	...
	Behavior	C	C	D	C	D	...
CTFT	Own standing	g	g	g	g	g	
	Other's standing	g	g	b	g	b	
	Payoff	$b-c$	$b-c$	b	$-c$	b	

(b) Pairs of Pavlov players recover from perception errors.

	Round	n	**X** $n+1$	$n+2$	$n+3$	$n+4$	
Pavlov	Payoff	$b-c$	$b-c$	$-c$	0	$b-c$	
	Behavior	C	**D**	C	D	C	...
Pavlov	Behavior	C	C	D	D	C	...
	Payoff	$b-c$	$b-c$	b	0	$b-c$	

(c) Pavlov is easily exploited by ALLD.

	Round	n	$n+1$	$n+2$	$n+3$	$n+4$	
Pavlov	Payoff	$-c$	0	$-c$	0	$-c$	
	Behavior	C	D	C	D	C	...
ALLD	Behavior	D	D	D	D	D	...
	Payoff	b	0	b	0	b	

like ALLD. Table 4.5(c) shows the behavior sequence that arises from pairing Pavlov with ALLD. Pavlov keeps trying to cooperate with ALLD as if ALLD's defections were mistakes. Thus while Pavlov does well when common, it cannot invade the population. CTFT, in contrast, has a much easier time invading, since it responds to ALLD just like TFT does. This suggest that when there are perception errors, it will be much harder to get cooperation started. However, Pavlov can invade once CTFT or a similar strategy has risen to high frequency. Thus some nice strategies may provide the environment necessary for others to evolve.

So what should we make of all this? We think that the Axelrod-Hamilton model shows that some reciprocating strategy is the likely evolutionary outcome when individuals interact many times. However, which reciprocating strategy evolves usually depends on the kind and frequency of errors that individuals make.

4.3 Partner choice

The Axelrod-Hamilton model assumes that individuals are paired at random. After that the length of the interaction varies from pair to pair but doesn't depend on the pair's behavior. This assumption strikes many students as deeply unrealistic. Wouldn't it make more sense to assume that cooperative relationships last longer than uncooperative ones? After all, people (and likely other social creatures) choose their friends and coalition partners—shouldn't this make cooperative outcomes more likely?

Allowing players to choose social partners can make a difference, sometimes an important one. However, first it's important to see that, when interpreted correctly, the basic Axelrod-Hamilton model already includes a form of partner choice. To see why, think about the model in the context of a group-living creature like baboons. These monkeys usually live in social groups of 20 to 60 individuals. There's lots of gene flow between groups, so two individuals who don't share an immediate kin connection are likely to be almost randomly sampled from the gene pool of the baboon population. (We say almost because there is some background relatedness, so that these two individuals are likely to be slightly related.) This means that you can think of each individual in the troop as having been paired with, say, 30 other individuals who are not known relatives. As long as the behavior in each of these relationships is independent, we can apply the Axelrod-Hamilton model to each one separately. So a reciprocator will establish cooperative relationships with other reciprocators but will quickly cease cooperating with the defectors. Thus, individuals choose their social partners by continuing to cooperate with some of the 30 they were initially paired with—a kind of partner choice.

However, this way of thinking about the Axelrod-Hamilton model fails to represent partner choice when:

1. The payoffs for mutual defection are different than the payoffs associated with not interacting. This is of particular importance when mutual defection has a lower payoff than simply not interacting.

2. Interactions within a pair affect other individuals. This can occur for many reasons. The simplest is when limited resources force individuals to choose whom to cooperate with. If Rob can only help one individual, and both Joe and Richard need help, Rob will have to make a choice.

In this section we consider a model of partner choice by Magnus Enquist and Olof Leimar.[46] This model applies when individuals are limited to one partner at a time. The model suggests, contrary to many people's intuitions, that partner choice actually *reduces* the range of conditions under which reciprocating strategies are evolutionarily stable.

Assume a large population in which each individual is paired at random with a single other individual. Each relationship continues each time period with probability w and thus these pairings last on average $1/(1-w) = R$ time periods. So far this is just like the Axelrod-Hamilton model. Now, however, when a relationship ceases, both partners search for a new partner. Their probability of finding one during each time period is $1-s$, and with probability s they must continue searching. This means that individuals search, on average, $1/(1-s) = S$ time periods before they find a new partner. Finally, the probability that individuals survive each time period is λ. Thus the average life span is $1/(1-\lambda) = L$. We assume that $L \gg S, R$ so that individuals have many relationships and many search periods during their lives.

Suppose that TFT is common in the population. Then the expected payoff of a TFT individual paired with another TFT individual is

$$V(TFT|TFT) = \frac{b-c}{1-w} = R(b-c).$$

After an individual's relationship breaks up, she has to search S time periods before she finds a new partner. Assume that the individual gets zero payoff during this period. Then the average payoff per period over the whole cycle is the payoff divided by the number of periods:

$$\frac{V(TFT|TFT)}{R+S} = \frac{R(b-c)}{R+S}.$$

Since TFT is common, a TFT individual will be paired with another TFT individual and the cycle continues. Then since individuals live L time periods, the lifetime fitness of TFT is

$$W(TFT) = \frac{R}{R+S}(b-c)L + w_0.$$

Next we calculate the fitness of rare invading ALLD individuals. Since TFT is common, these individuals are always

---

Box 4.4 Partner choice ESS when search is lengthy

 When searching for a new partner takes a very long time, we can simplify the ESS condition. Notice that as $S \to \infty$, the denominators on both sides of Condition 4.6 are approximately equal to S because both R and 1 are small enough compared to S to ignore. This gives us

$$\frac{R}{S}(b - c) > \frac{1}{S}b.$$

The factor $1/S$ cancels out of both sides, leaving the condition

$$R(b - c) > b.$$

Since $R = 1/(1 - w)$, this simplifies to the same condition for TFT to be an ESS as in the basic Axelrod-Hamilton model, $wb > c$.

---

paired with TFT and receive b during the first period. Then the cooperator breaks off the relationship, and both parties search for a new partner. The search takes S time periods, and then ALLD invaders are paired once again with a TFT individual. Thus the payoff per time period for ALLD individuals is $b/(1 + S)$ and their lifetime fitness becomes

$$W(ALLD) = \frac{1}{1 + S}bL + w_0.$$

 Thus TFT is an ESS if

$$\frac{R}{R + S}(b - c) > \frac{1}{1 + S}b. \tag{4.6}$$

To understand what this condition is telling us, let's look at the extreme values of S. First, notice that if it takes a long time to find a new partner compared to the time spent cooperating ($S \to \infty$), then this condition becomes $R(b - c) > b$, which is exactly the same as in the Axelrod-Hamilton model (Box 4.4). This makes sense because very long search times mean that, in effect, individuals only interact once—there is no partner choice.

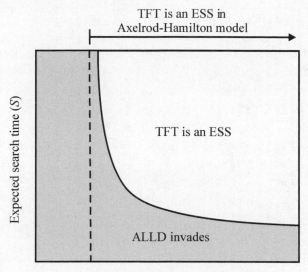

Figure 4.2: This plot shows the combinations of R and S that allow TFT to be an ESS. The vertical dashed line corresponds to $R = \frac{b}{b-c}$, the stability condition for TFT against ALLD in the Axelrod-Hamilton model. When relationships last longer than this value, TFT is an ESS in the absence of partner choice. This figure shows that allowing partner choice reduces the range of conditions under which reciprocity is stable because it permits defectors to exploit a series of reciprocators, especially when search times are short. (Redrawn from Enquist and Leimar 1993.)

Next, assume that $S = 0$ so that new partners are found immediately. Then the condition for TFT to be an ESS is $b - c > b$, which is never satisfied. When there are no search costs and TFT is common, defectors can victimize one reciprocator and then move to another and then another. Figure 4.2 plots the combinations of search time (S) and relationship time (R) that allow TFT to be evolutionarily stable. When search time is long, defectors get little advantage from partner choice. But as search time decreases, the range of conditions under which TFT is an ESS decreases.

4.4   Indirect Reciprocity

Everyday intuitions about reciprocity seem to be rooted in the notion of exchange. I help somebody, and in return, he helps me in the future. Reciprocal helping is not thought to be altruistic because it is repaid—over the entire course of actions, the payoffs are mutualistic, not altruistic. A similar *quid pro quo* view dominated evolutionary thinking about reciprocity for years, so much so that the term "return benefit" altruism was sometimes substituted for reciprocal altruism. This began to change in 1987 when Richard Alexander suggested that large-scale human cooperation was based on *indirect reciprocity* which "involves reputation and status, and results in everyone in the group continually being assessed and reassessed." The basic idea was that selection could favor the strategies that choose to help individuals who had helped others in the past. Many (including one of your authors) were skeptical about Alexander's proposal, but in the late 1990s Martin Nowak and Karl Sigmund showed that Alexander's argument was cogent. In this section, we first present a simplified version of the Nowak-Sigmund model without taking account errors. Then we show how errors modify the results.

4.4.1   Nowak-Sigmund model

Assume that there is a large population. During each time period, individuals meet in randomly selected pairs and interact in a prisoner's dilemma. As usual, if they cooperate, they produce a benefit b for the other individual at a cost c to themselves; if they defect, they produce nothing and experience no cost. Because the population is large, every pair is novel—actors never meet the same individual twice. The population continues to interact with probability w, so that on average individuals are paired $1/(1-w)$ times. Reproductive success is proportional to the sum of the payoffs. There are two strategies: Discriminator (DSC) cooperates on the first interaction and then cooperates on each subsequent interaction if the individual he is paired with cooperated on the previous round. The second strategy is ALLD.

It is easy to show that DSC is an ESS in this model. Let's compute the expected payoffs of the two strategies. When DSC is common, individuals using this strategy are paired again and again with another DSC individual. This means that they cooperate every period and achieve the payoff

$$W(DSC) = (b - c)(1 + w + w^2 + \ldots) = \frac{b - c}{1 - w}.$$

Rare ALLD individuals are also paired with a Discriminator each turn. Thus they defect during the first period and earn b. During subsequent periods, they earn nothing because Discriminators don't help individuals on the previous round. This earns rare ALLD a fitness

$$W(ALLD) = b.$$

But these are exactly the same payoffs as TFT against ALLD in the Axelrod-Hamilton model! And this means that DSC is an ESS in exactly the same conditions as TFT even though individuals never interact with the same person. What gives?

The answer is that what really counts is knowledge about past behavior. In both direct and indirect reciprocity, individuals know the previous behavior of others, and this allows contingent strategies to cooperate with other cooperators and defect against defectors, which in turn gives rise to the assortative interaction that is required if cooperation is to be favored by natural selection. At least as far as ESS conditions go, it doesn't really matter whether you meet the same individual each period or a different individual each period, as long as you know the history of all your partners.

4.4.2   Mistakes and indirect reciprocity

Mistakes play a big role in indirect reciprocity, even bigger than their role in direct reciprocity. ("Above the title" as they say in Hollywood.) First we examine why Discriminator can't be an ESS when there are errors. Then we discuss alternative strategies that are more robust.

As before, we assume that individuals who mean to cooperate occasionally defect by mistake. Individuals know that they have defected, so these are *implementation errors*. We assume that errors are rare (exactly how rare we'll see in a minute) and that individuals never cooperate by mistake. Now let's see what happens when a mistake occurs. Discriminators are common, but there are a few ALLC individuals in the population.

First focus on the fitness of a Discriminator (DSC) who makes a mistake during his interaction with a series of other Discriminators (Table 4.6(a)). As long as errors are rare, during the next turn he will interact with a DSC individual who cooperated on the previous turn. Our focal Discriminator will cooperate, but the other DSC will not. Luckily for DSC, this outcome is the same for an ALLC individual interacting with a series of Discriminators. If you consider in Table 4.6(a) how a focal ALLC will behave, you will see that the outcome will be identical. Provided we consider an implementation error by a focal individual, committing an error has no effect on the relative fitness of DSC and ALLC. And because rare errors create short deviations from all cooperation, we don't need to calculate expected fitness for the entire interaction— only compare the effects of an error. ALLC will not invade, as a result of errors hurting those who commit them.

Things change, however, when we consider what happens to a Discriminator whose current partner made an error on the previous turn. In this case, ALLC does better than DSC. Here is why. Consider a focal DSC individual first (Table 4.6(b)). If his partner made an error on the previous move, he has to defect, but this means that during the next turn he will be not be helped by the next DSC he is paired with. So he gets b on the turn he doesn't help (to punish his current partner's defection on the previous round) and $-c$ on the next turn when he helps but is punished for his previous

Table 4.6: Implementation errors and the evolution of indirect reciprocity when Discriminator (DSC) is common. Each table shows a sequence of interactions between a focal individual and a series of DSC partners, a new one each round. Paired behaviors are marked in boldface. Defections resulting from mistakes are marked in parentheses. (a) A focal DSC who makes an error gets the same payoff as a focal ALLC making the same error. Both are punished once. (b) A focal DSC who punishes a mistake is also punished by the next DSC he interacts with. (c) A focal ALLC never punishes a mistake, so is not punished by the next DSC. Thus ALLC does better than DSC when implementation errors are possible.

(a) Focal Discriminator makes an error and suffers the same consequences as ALLC would.

	Round	n	$n+1$	$n+2$	$n+3$	...
Focal DSC	Payoff	$b-c$	b	$-c$	$b-c$	...
	Behavior	**C**	**(D)**	**C**	**C**	...
DSC n	Behavior	**C**	C	C	C	...
DSC $n+1$	Behavior	C	**C**	C	C	...
DSC $n+2$	Behavior	C	C	**D**	C	...
DSC $n+3$	Behavior	C	C	C	**C**	...

(b) Focal Discriminator punishes a mistake and suffers for it.

	Round	n	$n+1$	$n+2$	$n+3$	...
Focal DSC	Payoff	$b-c$	b	$-c$	$b-c$	...
	Behavior	**C**	**D**	**C**	**C**	...
DSC n	Behavior	**C**	C	C	C	...
DSC $n+1$	Behavior	(D)	**C**	C	C	...
DSC $n+2$	Behavior	C	C	**D**	C	...
DSC $n+3$	Behavior	C	C	C	**C**	...

(c) Focal ALLC never punishes a mistake.

	Round	n	$n+1$	$n+2$	$n+3$	...
Focal ALLC	Payoff	$b-c$	$b-c$	$b-c$	$b-c$	...
	Behavior	**C**	**C**	**C**	**C**	...
DSC n	Behavior	**C**	C	C	C	...
DSC $n+1$	Behavior	(D)	**C**	C	C	...
DSC $n+2$	Behavior	C	C	**C**	C	...
DSC $n+3$	Behavior	C	C	C	**C**	...

defection. Thus the average contribution to his fitness of this sequence is $b - wc$ (remember that there is still only a chance w that he reaches the second round, where he gets $-c$).

Now compare this outcome to an ALLC individual whose current partner defected on the previous round (Table 4.6(c)). Tolerant (or lazy) ALLC ignores the defection and cooperates, and thus receives the benefits of cooperation during the next period, and receives a payoff of $b - c$ each period. The average contribution to his fitness of this sequence is $(b - c) + w(b - c) = (1 + w)(b - c)$. Thus ALLC invades whenever

$$(1 + w)(b - c) > b - wc$$
$$wb > c.$$

ALLC will invade because $wb > c$ is also the condition for reciprocity to be ESS against ALLD. ALLC can invade whenever ALLD cannot. Either DSC is replaced by ALLD or, if implementation errors occur, DSC is replaced by ALLC. Errors put DSC at a disadvantage because Discriminators punish errors by refusing to cooperate. Such police work is necessary to keep defectors at bay. The problem is that doing police work causes Discriminators to be punished on the next turn.

This problem can be solved by strategies that only punish "unjustified" defections. Consider the strategy Reputation Discriminator (RDSC) that uses a standing mechanism similar to that of Contrite-Tit-for-Tat (CTFT). RDSC cooperates with individuals in good standing or whose standing is unknown and doesn't cooperate with those who are not in good standing. Initially, standing is unknown. Individuals remain in good standing if they cooperated in the previous round or if they defected against a partner who was not in good standing. Defecting against someone in good standing results in bad personal standing. In this way, RDSC distinguishes between justified defections that occur when punishing a defection and unjustified defections against a partner in good standing.

Table 4.7: Implementation errors and the evolution of indirect reciprocity when Reputation Discriminator (RDSC) is common. The table shows a sequence of interactions between a focal individual and a series of RDSC partners, a new one each round. Each row shows the behavior in that round, followed by the standing the individual had at the start of that same round. Paired behaviors are marked in boldface. Defections resulting from mistakes are marked in parentheses. The individual who made an implementation error in round n gains bad standing for round $n+1$ and is punished by the focal RDSC. The focal individual, however, retains good standing for round $n+2$ because defecting against an individual in bad standing is justified. Thus, unlike DSC, the focal individual enjoys mutual cooperation in round $n+2$.

	Round	n	$n+1$	$n+2$	$n+3$	...
Focal RDSC	Payoff	$b-c$	b	$b-c$	$b-c$	...
	Beh/Stand	**C/g**	**D/g**	**C/g**	**C/g**	...
RDSC n	Beh/Stand	**C/g**	C/g	C/g	C/g	...
RDSC $n+1$	Beh/Stand	(D)/g	**C/b**	C/g	C/g	...
RDSC $n+2$	Beh/Stand	C/g	C/g	**C/g**	C/g	...
RDSC $n+3$	Beh/Stand	C/g	C/g	C/g	**C/g**	...

RDSC excludes defecting strategies like ALLD in the same way as DSC. The ESS conditions are exactly the same. However, RDSC can also resist invasion by ALLC. To see why, let's compare what happens when a Reputation Discriminator's partner makes a mistake to what happens to ALLC in the same situation (Table 4.7). When the focal RDSC individual's partner has made an error in the previous round, she is bad standing the next period when she is interacting with the focal. The focal RDSC defects but remains in good standing, and therefore the next period his new partner cooperates.

This means the the fitness increment for RDSC is $b+w(b-c)$. The payoff to ALLC is exactly the same as we computed before (see Table 4.6(c)), $(1 + w)(b - c)$. Thus, RDSC has a higher payoff than ALLC. It can avoid one episode of costly cooperation without any loss to its reputation. In effect, punishment of defectors pays. Thus standing-based strategies like RDSC can be evolutionarily stable when implementation errors are possible, while pure-behavior strategies like DSC probably cannot.

However, this kind of strategy comes with a cost. Discriminators only had to know what others did on the previous turn. Reputation Discriminators have to know whether what others did was justified by the circumstances. It seems to us that this is the real difference between direct and indirect reciprocity. In both cases, dealing with mistakes requires detailed knowledge about past behavior and whether that behavior was justified given the circumstances. It is not much of a stretch to assume that people (or baboons) have good knowledge about individuals with whom they have a long-term relationship, and that's all that's needed to make direct reciprocity work. However, to make indirect reciprocity work, individuals have to have the same quality of knowledge about the behavior of everybody in the population. This seems much less plausible.

4.5   Reciprocity and collective action

All of the models of reciprocity have assumed that cooperation occurs in pairs. However, many, many cooperation problems involve larger numbers of unrelated individuals. For example, individuals might agree to cooperate in mutual defense of their territory. Each risks life and limb for the benefit of not just one other person, but many. Such *collective action* problems have the same logic as the prisoner's dilemma—increasing the number of cooperators increases the average group payoff, and an individual who switches from cooperation to defection is always better off. We will model collective action problems using a simple linear version of the n-person prisoner's dilemma game in which each individual who cooperates generates a benefit b that is shared equally among the n members of the group. Each individual who cooperates pays a private cost c. For this to be a PD, $b - c > c > b/n$.[47]

Collective action problems often recur repeatedly in social groups, suggesting an iterated structure similar to what we have worked in up to this point. Thus it is easy to believe that the lessons learned in the dyadic prisoner's dilemma also apply to the n-person game—contingent strategies that cooperate only when others cooperate will lead to the evolution of cooperation. In this section we will see that this is only half right. Contingent strategies can be evolutionarily stable in repeated collective action problems. However, small amounts of relatedness do not allow such strategies to increase when rare. In fact, there is no obvious mechanism that will reliably destabilize defecting equilibria except selection among groups. We begin by analyzing the repeated n-person prisoner's dilemma in which reciprocators punish defectors by defecting. We will see that this is not a very good strategy unless groups are quite small. We will then analyze models in which punishment takes other forms, and this leads to more effective strategies.

The methodological payoff in this section comes from seeing how to model n-person games in which social groups are larger than pairs and the number of possible group compositions explodes. As we will see, the core logic is identical to the two-person case. The complexity is only superficial.

4.5.1 n-person reciprocity

Assume that there is a large population. Each generation, individuals are sampled randomly into groups of size n. Each individual in each group of n individuals can cooperate or defect each round. Cooperation generates a benefit b shared equally by everybody in the group as a whole at a cost c, where $b - c > c > b/n$. Again, assume a chance w each round that the group persists for another round and interacts again.

We define a family of reciprocating strategies T_i as follows: T_i always cooperates on the first round. On subsequent rounds, T_i cooperates only when at least i other individuals $(0 \le i \le n-1)$ cooperated on the previous round. Here we will focus on the least tolerant member of the family, T_{n-1}, who only cooperates if every other individual in the group cooperated in the previous round. In the problem set for this chapter, you will study other members of the family.

Let's compare this TFT-like strategy with TFT. First, let's find out when T_{n-1} is stable against ALLD. In order to write payoffs in these n-person games, we'll need to add a bit of notation. Define $V(T_{n-1}|x)$ as the payoff to a T_{n-1} individual, given that x of the *other* individuals in the group have the strategy T_{n-1}. Thus when T_{n-1} is common, the payoff to a T_{n-1} individual is

$$V(T_{n-1}|n-1) = \frac{nb}{n} - c + w\left(\frac{nb}{n} - c\right) + w^2\left(\frac{nb}{n} - c\right) + \ldots$$
$$= \frac{b-c}{1-w}.$$

Each cooperates, and since no one ever defects, cooperation continues for $1/(1-w)$ rounds, as in the pairwise game. The payoff to an ALLD individual invading such a group is

$$V(ALLD|n-1) = \frac{n-1}{n}b$$

because ALLD receives the benefits of $n-1$ individuals, shared among n group members, but then no one ever cooperates again. Thus T_{n-1} is stable against ALLD when

$$\frac{b-c}{1-w} > \frac{n-1}{n}b.$$

This condition says that T_{n-1} is stable against ALLD when the long run advantage of cooperation (left side) is greater than the net cost of cooperation (right side). This is a lot like the condition for TFT to resist ALLD. Notice that when groups are very large, the right-hand side of this expression is approximately b, the same result as for the dyadic model.

It looks like the result in the n-person case is qualitatively similar to the pairwise case.

But what is much harder here is getting reciprocity started. TFT could not invade ALLD, but with a little assortment within pairs, provided by kinship, for example, TFT could readily invade. As groups become larger, however, it gets harder and harder to get enough reciprocators together to get the system started. With T_{n-1}, the only kind of group in which this strategy does better than ALLD is when everyone is T_{n-1}. How hard is it to get a group of all T_{n-1}? Let p be the frequency of T_{n-1} in the metapopulation. Then if groups are formed at random, the chance that a given T_{n-1} finds herself in a group comprised entirely of such individuals is p^{n-1}. Unless reciprocators are very common or groups are very small, the probability will be miniscule.

You might think that kinship will save the day, as it did in the dyadic invasion case. It turns out that it doesn't help much here. Using the haploid kin-selection model again, let's calculate expected fitness of a rare T_{n-1} invader. There are two kinds of groups we need to consider: those with fewer than $n - 1$ *other* T_{n-1} and those with exactly $n - 1$ other T_{n-1}. First, suppose that there are j other T_{n-1} individuals in his group, where $j < n - 1$. Then all T_{n-1} individuals cooperate on the first turn and defect during all remaining turns. So, the expected fitness of the individual is

$$V(T_{n-1}|j) = \frac{j+1}{n}b - c + w(0) + w^2(0) + \ldots$$
$$= \frac{j+1}{n}b - c.$$

Second, if there are $n - 1$ other T_{n-1} individuals, they all cooperate for as long as the group persists. Thus

$$V(T_{n-1}|n - 1) = (b - c)(1 + w + w^2 + \ldots) = \frac{b - c}{1 - w}.$$

To compute the expected fitness of T_{n-1}, we need to average over all of the different group compositions weighted by the probability that they occur:

$$W(T_{n-1}) = w_0 + \sum_{j=0}^{n-1} V(T_{n-1}|j) \Pr(j|T_{n-1}),$$

where $\Pr(j|T_{n-1})$ is the probability that each other member of the group is T_{n-1} given that the focal individual is T_{n-1}. To combine these expressions, notice that the fitness of an individual in a group with $n-1$ other T_{n-1} individuals can be written

$$V(T_{n-1}|n-1) = \frac{(n-1)+1}{n}(b-c) + w\frac{b-c}{1-w}.$$

Using this expression, we obtain for the overall expected fitness of T_{n-1}

$$W(T_{n-1}) = w_0 + \sum_{j=0}^{n-1} \Pr(j|T_{n-1})\left(\frac{j+1}{n}b - c\right)$$
$$+ \Pr(n-1|T_{n-1})w\frac{b-c}{1-w}.$$

The first term gives the expected payoff during the first interaction over all kinds of groups, and the second term gives the long run payoff in those groups with n total reciprocators. In Box 4.5, we show how to prove that

$$W(T_{n-1}) = w_0 + \left(\frac{E(j)+1}{n}b - c\right) + \Pr(n-1|T_{n-1})w\frac{b-c}{1-w}.$$

Remember now that we are asking when T_{n-1} can invade. Using the haploid model of kinship (Chapter 3), we find the expected number of reciprocators among the other $n-1$ individuals to be $E(j) = (n-1)r$ (see Box 4.6). The probability that all to be $n-1$ are reciprocators is r^{n-1} (again see Box 4.6). Thus the expected fitness of rare T_{n-1} invaders is

$$W(T_{n-1}) = w_0 + \underbrace{\left(\frac{(n-1)r+1}{n}\right)b - c}_{A} + \underbrace{r^{n-1}w\frac{b-c}{1-w}}_{B}.$$

Box 4.5 Taking the expectation across groups
The only term presenting difficulty is

$$\sum_{j=0}^{n-1} \Pr(j|T_{n-1}) \left(\frac{b}{n}(j+1) - c \right).$$

Sums like this can be decomposed across addition, yielding

$$\frac{b}{n} \left(\sum_{j=0}^{n-1} \Pr(j|T_{n-1})j + \sum_{j=0}^{n-1} \Pr(j|T_{n-1})(1) \right) - \sum_{j=0}^{n-1} \Pr(j|T_{n-1})c.$$

Since the probability being summed up is the entire distribution of group compositions, $\sum_j \Pr(j|T_{n-1}) = 1$ and $\sum_j \Pr(j|T_{n-1})j = \mathrm{E}(j)$. This gives us

$$\frac{b}{n} \left(\mathrm{E}(j) + 1 \right) - (1)c = \frac{\mathrm{E}(j)+1}{n}b - c.$$

The expectation of the number of other T_{n-1} individuals, $\mathrm{E}(j)$, depends upon what model of group formation we assume.

All of this must exceed $W(ALLD) = w_0$ in order for T_{n-1} to invade. The term marked **A** gives the inclusive fitness from cooperation during the first interaction, and **B** gives the expected benefit from long run reciprocal cooperation. If **A** is positive, then cooperation can evolve purely for the direct inclusive fitness benefits. Reciprocity is not needed. So imagine for the moment that this term is negative. Now notice that the term marked **B** is proportional to r^{n-1}, which rapidly approaches zero as group size increases, even if r is very large. For example, suppose $r = 0.5$ (full siblings). Then for $n-1 = 10$, $r^{n-1} \approx 10^{-3}$, and for $n-1 = 20$, $r^{n-1} \approx 10^{-6}$. This means that the synergism between reciprocity and kinship disappears as soon as groups exceed a handful of individuals. If the first term is positive, cooperation is favored by kin selection and reciprocity is irrelevant. If the first term is negative, reciprocity will not increase even if individuals are closely related.

Box 4.6 n-person kin-biased group formation

In Chapter 3, we showed that under kin-biased group formation, the probability of sampling an individual of type x, given that the focal already in the group is of type x, is

$$\Pr(x|x) = r + (1 - r)p$$

where r is the coefficient of relatedness between the individuals and p is the frequency of type x in the population from which the individuals are sampled. In this case, $p \approx 0$, because we are asking when T_{n-1} will invade. This means $\Pr(T_{n-1}|T_{n-1}) \approx r$. Thus if we sample $n - 1$ other individuals, each with this probability of being T_{n-1}, the expected number of T_{n-1} individuals will be $(n - 1)r = \mathrm{E}(j)$ in the fitness expression $W(T_{n-1})$. Similarly, $\Pr(n - 1|T_{n-1}) = r^{n-1}$.

You might think also that making the reciprocators more forgiving, maybe by using a strategy like T_{n-2} or T_{n-3}, would help altruism in the n-person case. It turns out that T_{n-1} is the only ESS, as you will show when you solve Problem 6.

4.5.2   Altruistic punishment

Humans sometimes cooperate in large groups of unrelated individuals. If reciprocity is a nonstarter for explaining these cases, what is left? A number of biologists and economists have begun discussing altruistic punishment, a combination of altruism and punishment of nonaltruists. Trivers called this sort of behavior *moralistic reciprocity*.[48] The trouble with direct reciprocity in large groups, as we saw, is that an individual cannot withhold cooperation from defectors without simultaneously punishing other cooperators. A second punishment stage can solve this problem, since punishment can be directed to defectors even when benefits cannot be denied to them. However, you will see that this solution creates two additional problems.

Imagine a population which consists of a large number of social groups, each of size n. Groups are formed at random each generation and persist with probability w each round. In each round, all members play an n-person PD, as defined before. Immediately after the PD in each round, each player has the opportunity to punish any other player at a cost k. A punished individual's payoff is reduced by an amount h (harm inflicted). Suppose two strategies for this world:

Altruistic Punishers (AP): Always cooperate and always punish any individual who did not cooperate in the most recent PD phase.

Reluctant Cooperators (RC): Always defect, until punished. After being punished, cooperate for the rest of the generation.

The reason for assuming RC instead of ALLD is that in a population of AP, ALLD will always do worse than RC because ALLD never catches on to the fact that it is being punished. RC instead recognizes when defection is costly and begins cooperating. Thus RC will always replace ALLD in any population with AP present.

To see what happens in this system, we need to write payoffs to each strategy given a certain group composition, as we discussed. The round one payoff to AP in a group with x other APs is

$$V_1(AP|x) = (x+1)\frac{b}{n} - c - (n-x-1)k.$$

A fraction $(x+1)/n$ of the group cooperates on the first move. Then AP punishes any individuals who did not cooperate, which is the total group size minus itself and minus any other APs. In any round afterwards, all RCs have been coaxed into cooperating, so the AP individual enjoys cooperation for the remaining duration of the generation. This yields for the total payoff

$$V(AP|x) = (x+1)\frac{b}{n} - c - (n-x-1)k + \frac{w}{1-w}(b-c).$$

The payoff to a RC individual is similarly

$$V(RC|x) = \begin{cases} x > 0, & xb/n - xh + \frac{w}{1-w}(b-c) \\ x = 0, & 0 \end{cases}.$$

Let's derive the invasion criteria. AP is an ESS against RC when

$$V(AP|n-1) > V(RC|n-1)$$

$$(b-c)\left(1 + \frac{w}{1-w}\right) > (n-1)\frac{b}{n} - (n-1)k + \frac{w}{1-w}(b-c).$$

This simplifies quite quickly to yield

$$(n-1)h > c - \frac{b}{n}.$$

If the cost of being punished by the rest of the group is greater than the net cost of cooperation, then AP is an ESS against RC. Note that the cost of punishing (k) is irrelevant for this condition. Even if punishment is quite costly, all that matters is the cost inflicted on others. When can AP invade RC?

$$V(AP|0) > V(RC|0)$$

$$\frac{b}{n} - c - (n-1)k + \frac{w}{1-w}(b-c) > 0.$$

Again, this simplifies very quickly:

$$w\frac{b-c}{1-w} - (n-1)k > c - \frac{b}{n}.$$

If the benefit of long-term cooperation minus the cost of punishing everyone else (the left side) exceeds the cost of cooperation (the right side), then AP can invade. Thus groups must persist for a long time, or punishment must be quite cheap. Note, however, that this condition is much easier to satisfy than the requirements for direct reciprocity in large groups that we analyzed before.

4.5.3   Who watches the watchmen?

The model so far is seriously incomplete, however. We have analyzed first-order defection, the choice between cooperation and noncooperation. By introducing punishment, however, we have created a second-order cooperation dilemma. What happens if some individuals cooperate but do not punish? Provided other individuals pay the cost of punishment, second-order free riders of this kind will enjoy the benefits of prolonged cooperation without paying the costs of policing. Every watchman, so to speak, has the incentive to slack and let some other watchman produce the public good of first-order cooperation.

To grasp the severity of this dilemma, let's see what happens when we introduce ALLC into the mix. The payoff to ALLC in a group without RCs is always

$$V(ALLC|n-1) = \frac{b-c}{1-w} = V(AP|n-1).$$

Thus ALLC can always drift into a population of AP. If enough ALLCs drift in, they will eventually allow RCs to invade. Deriving the critical frequency of AP to keep out RC isn't particularly difficult, but it is tedious. The key insight is that we need some mechanism to keep the frequency of AP high.

One of your authors (R. B.) has done a lot of work exploring (or has a unhealthy obsession with) solutions to the second-order dilemma in large groups. One solution is to modify the AP strategy such that it punishes not only defectors, but also any individual who fails to punish.[49] In the presence of rare errors, ALLC will suffer more than AP, and so this recursive rule will favor AP. This solution is unsatisfactory, however, because it seems like people are incapable of calculating large numbers of recursive steps. Is it really plausible that people remember whether one punished the nonpunisher of the nonpunisher of the nonpunisher? Another solution is group selection.[50] Another is conformist learning.[51] Yet both of these solutions depend upon cultural, rather than genetic, transmission in order to be plausible. This may explain why cooperation and punishment in large groups of unrelated individuals appears to be confined to human societies. Finally, collective action can be linked to a system of indirect reciprocity so that those who fail to contribute to collective action are punished by being denied help in dyadic interactions regulated by indirect reciprocity. Like the RDSC strategy discussed before, such punishment benefits the punisher and thus does not suffer the second-order free rider problem.[52]

Guide to the Literature

Iterated prisoner's dilemma. It all begins with Triver's 1971 seminal paper, which laid out the problem and, in an appendix, suggested that reciprocity could be formalized using a repeated prisoner's dilemma. It wasn't until 1981 that Axelrod and Hamilton published their analysis of the evolution of reciprocity described in this chapter. This led to an enormous number of theoretical papers in a wide range of disciplines. The first decade or so was reviewed by Axelrod and Dion 1988. It is worth noting that many of the same ideas were developed independently by game theorists in economics. The work of Reinhardt Selten 1975 is of particular importance. There has been much less empirical work published—the classic examples are Wilkinson's 1984 work on blood sharing in vampire bats, Seyfarth and Cheney's 1984 field experiments with vervet monkeys, and the (often-debated) laboratory work of Milinski 1987 on predator inspection in sticklebacks. **Errors and standing.** One of the earliest analyses of standing-based strategies is in Sugden's 1986 book, which helped reintroduce evolutionary game theory to economics. See Gintis 2000 for a textbook treatment of the important role of evolutionary game theory in economics today. Further work on standing strategies can be found in Boyd 1989, Boerlijst et al. 1997 and especially

Leimar 1997. Nowak and Sigmund have written extensively about strategies designed to deal with perception errors. See Nowak and Sigmund 1993. The classic works on partner choice are Enquist and Leimar 1993 and Dugatkin and Wilson 1991. Batali and Kitcher 1995 dealt with the case in which nonparticipants have a lower payoff. **Indirect reciprocity** was first introduced by Alexander 1987 in his book the *Biology of Moral Systems*. Boyd and Richerson 1989 interpreted Alexander, but the first really successful model was published by Nowak and Sigmund 1998a; 1998b. Critiques of this work can be found in Leimar and Hammerstein 2001 and Panchanathan and Boyd 2003. Nowak and Sigmund 2005 provide a useful review of recent developments. **Collective action.** The early work on collective action can be found in Taylor 1976 and Axelrod 1986. Boyd and Richerson 1988 and Joshi 1987 published very similar analyses of the repeated n-person PD about the same time. The evolution of altruistic punishment is modeled by Boyd and Richerson 1992, Henrich and Boyd 2001, Boyd et al. 2003, and Panchanathan and Boyd 2004. For reviews of empirical work on punishment, see Clutton-Brock and Parker 1995 for other animals and Fehr and Gächter 2002 for humans.

Problems

1. Tiny basins. In the first part of this chapter we saw that in a population in which only ALLD and TFT compete, both strategies are evolutionarily stable once common. Show that TFT will increase in frequency as long as its frequency exceeds the threshold value

$$\hat{p} = \frac{c(1-w)}{w(b-c)} = \frac{1-w}{w(b/c - 1)}.$$

What happens to the basin of attraction of ALLD as w increases?

2. Even bad guys are vulnerable. Show that a population in which ALLD is common can be invaded by STFT when mutation maintains TF2T in the population at a low frequency.

3. More tolerance. Some people have explored the evolution of strategies which are more tolerant than TFT to defections. One example is Tit-for-Two-Tats (TF2T), which begins by cooperating and only defects if its opponent defected on the two previous rounds. (a) Show that the fitness of TF2T in a population of TF2T and ALLD is

$$W(TF2T) = p\frac{b-c}{1-w} - (1-p)(1+w)c + w_0$$

where p is the frequency of TF2T. (b) Show that TF2T is an ESS against ALLD when

$$w^2b > c.$$

Compare this result to the condition for TFT to be an ESS against ALLD. What do these results suggest about the evolution of more-tolerant strategies? How might these results be sensitive to the presence or absence of errors? (c) Show that the condition for any strategy TFnT that waits n defections before retaliating with defection to be an ESS against ALLD is

$$w^nb > c.$$

4. Doing the dishes again, and again. Recall the problem from Chapter 2 about doing the dishes. Assume the same model, but now the game is iterated such that pairs go on to another interaction with probability w. Previously, a stable mix of individuals who wash (cooperators) and do not wash (defectors) was stable. In this problem, you will prove that iteration can allow cooperators to exclude defectors in this game as well. Imagine a strategy called Wash-for-Tat (WFT), which alternates washing with its opponent, but washes the dishes only if its opponent washed on the last round. If two WFTs meet, one of them washes on the first round at random (they flip a coin). If WFT meets a strategy which insists on washing the first round, WFT does not wash on the first round. If the other strategy refuses to wash on the first round, WFT washes. On all subsequent rounds, WFT behaves as its opponent did on the previous round. Thus two WFTs take turns washing the dishes. (a) Show that

$$V(WFT|WFT) = V(AW|AW) = \frac{b - k/2}{1 - w},$$

where AW is Always Wash, the strategy which always washes, regardless of what its opponent does. (b) Show that WFT is stable against invasion by rare NW when

$$wb > k/2.$$

Compare this result to Hamilton's rule.

5. Sequential reciprocity. Some people have argued that the simultaneous iterated prisoner's dilemma we analyzed in this chapter is a poor model of reciprocal altruism because most altruism of this type in nature is likely to be sequential rather than simultaneous. For example, if you take me to the airport, obviously I cannot simultaneously take you to the airport. Instead, we take turns receiving the benefit and paying the cost.

Construct a sequential model of reciprocal altruism in a world with only ALLD and TFT. In each turn of this model, only one player has the opportunity to help. TFT still cooperates when it goes first and otherwise behaves however its opponent did on its opponent's last move. Which player has the opportunity to help alternates each turn, and assume that which player begins with the opportunity to help is random. Show that this model has all the same equilibria and dynamics as the simultaneous model.

6. Tolerance and n-person reciprocity. We saw in this chapter that T_{n-1} is an ESS in the n-person IPD, much as TFT is an ESS in the two-person IPD. Show that any strategy T_{n-m}, where $m > 1$, is never an ESS in the same game.

7. Nice guys finish last. The standard model of the IPD seems to suggest that being nice is favored by natural selection. Robert Axelrod has stressed this finding: that strategies which begin by trusting their opponents and cooperating outperform those which are suspicious. Here you will prove that Axelrod's results rely upon the assumption that all w's are the same. If different pairs sometimes have different chances of persisting, it may pay to be nasty in the beginning. Assume the iterated PD as presented in the chapter. Now, however, there is a chance v that a pair has a chance w of interacting again each round. There is a $1 - v$ chance that the interaction instead lasts only one round ($w = 0$ for this pair). (a) Show that the payoff to a TFT against another TFT is now

$$V(TFT|TFT) = (b - c)\frac{1 - w(1 - v)}{1 - w}.$$

(b) Show that TFT resists invasion by ALLD when

$$v\frac{w(b - c)}{1 - w} > c.$$

(c) Define a strategy Hesitant Tit-for-Tat (HTFT). HTFT defects on the first round but cooperates on the second. On any turn after the second, HTFT plays as its opponent did on the previous round. Show that HTFT resists invasion by ALLD when

$$wb > c.$$

(d) Interpret your results from (b) and (c) with respect to which strategy does better in this environment.

8. n-person stag hunting. In the last chapter you analyzed evolution in the two-person stag hunt. Now assume that instead of pairs, individuals are sorted into groups of n. There are two strategies, as before, hunting hare (H) and hunting stag (S). Hunting hare is a solitary activity and always yields a small payoff h. Hunting stag is a cooperative behavior. Hunters attempt to surround the stag. If there aren't enough of them, the stag can escape and they get nothing. Thus assume that the chance a group of stag hunters captures a stag is

$$\frac{x}{n}$$

where x is the number of stag hunters in the group and n is the size of the group. If they capture a stag, each member of the hunting party gets a payoff $s > h$. (a) Calculate the conditions for H and S to be ESSs. (b) Let p be the frequency of S in the population as a whole. Find any internal equilibria and determine their stability.

Notes

[40]Seyfarth and Cheney 1984.

[41]This was first pointed out by Reinhard Selten and Peter Hammerstein in 1984.

[42]In finite populations gene frequencies change even when there is no selection because there will be chance variations in who reproduces and who doesn't. This process, called genetic drift, leads to aimless, random fluctuations in gene frequencies. See Rice 2004 for an introduction to modeling genetic drift.

[43]Boyd and Lorberbaum 1987.

[44]Sugden 1986, Boyd 1989.

[45]Nowak and Sigmund 1993.

[46]Enquist and Leimar 1983.

[47]It is possible to define a more general version of the n-person PD. See, for example, Boyd and Richerson 1988.

[48]Trivers 1971.

[49]Boyd and Richerson 1992.

[50]Boyd et al. 2003.

[51]Boyd and Richerson 1985, Henrich and Boyd 2001.

[52]Panchanathan and Boyd 2004.

Box 4.7 Symbols used in Chapter 4

b Fitness benefit of receiving altruism

c Fitness cost of providing altruism

w Probability of an interaction continuing for another turn

R Average length of a pairing (in partner choice model)

S Average length of search

L Average lifespan

n Number of individuals per social group

k Cost of punishing a defector

h Cost of being punished

Chapter 5

Animal Communication

The basic premise of game theory is that social life is like a giant game. If this is true, it is also true that life is less like chess and more like poker. In games like chess, information is entirely public. Each player knows which pieces are on the board, where they are positioned, and what moves are possible. In games like poker, however, there is private information, and access to this information strongly determines the outcomes of contests. Even when games are partly cooperative, like bridge, private information is crucial because cooperating players can benefit if they can find some way to communicate their private information to one another.

Communication is at least as important to animals as it is to card players. In confrontations, for example, if a stronger individual can convince its opponent that it will lose a fight between them, the stronger may win without a costly battle. The same goes for partly cooperative interactions. An offspring who is especially needy could benefit if it can communicate this fact to its mother. The problem, of course, is that individuals could also benefit by lying. If an ordinary individual can bluff its opponent into retreat, it also gets the resource without a fight, and the healthy but greedy

offspring can dupe its mother into giving it an extra morsel
of food.

There are many kinds of communication. Information can
be shared intentionally or accidentally, be inherently unfakeable or strategic.[53] In this chapter, we deal with two cases.
First, we present models in which intentional communication,
signals, can affect the outcomes of a game. Like much of the
literature in this area, we spend most of our time on models of
how the costs of producing a signal can affect how other individuals respond. Second, we treat explicitly social learning as
a special case of animal communication in which information
arises whether the demonstrator intends it or not—what is
sometimes called *social* or *public information*. Along the way,
we'll learn how to deal with two-locus genetic models and see
one way to model the evolutionary dynamics of information
within a population.

5.1   Costly signaling theory

Early animal behavior giants like Niko Tinbergen thought
that nature was full of cheap, honest signals, and that selection favored low-cost, honest communication.[54] This was the
dominant view of animal signals until 1978, when Richard
Dawkins and John Krebs[55] published a paper arguing that
honest signals will be rare because different animals never
share exactly the same interests. Instead, they argued that
signals are the product of an arms race between manipu-
lative signalers and suspicious receivers. This paper was
very influential, and many people came to believe that signals would rarely be honest and that early ethologists had
been misled by a failure to think clearly about individual
benefit.

A few years earlier, in 1975, an Israeli biologist named
Amotz Zahavi had published a paper in *Journal of Theoretical Biology* in which he argued that exaggerated characters in males, like peacock tails, could act as honest signals

of male quality because they were "handicaps" which only high-quality males could afford. This paper was controversial. Dawkins and Krebs argued against it in their 1978 paper, as had John Maynard Smith a few years earlier.[56] The problem was that few biologists could see how a signal could be honest just because it was costly. Why couldn't more frugal individuals save up resources and outreproduce the loudmouths? Since Zahavi's argument was nonmathematical and somewhat hard to follow, it was hard to be sure who was right.

This uncertainty persisted for some time in biology. Economists were far ahead of biologists in this because they started making models of these questions in the 1970s. Economists have long understood, for example, that signals can sometimes be honest because they are costly to produce. Thorstein Veblen argued for such a mechanism in his 1899 book *The Theory of the Leisure Class*, in which he argued that expensive luxury goods are unfakeable signals of wealth precisely because they are expensive, and in the early 1970's, several economists showed mathematically that his ideas were cogent. In 1977, the economist Jack Hirshleifer suggested that biologists might profit from borrowing costly signaling theory from economics.[57] He was right. The biologists Alan Grafen, H. C. Godfray, and Andrew Pominakowski later showed that Zahavi's idea could work.

In the sections that follow, we'll see examples of models in which costly signaling theory is put to the test. First, we will study the Sir Philip Sidney game, a very simple model which allows us to see how costly signaling theory works. Then we will consider a more realistic verison of the game that allows us to discuss some more subtle aspects of contemporary signaling theory.

5.1.1 The Sir Philip Sidney game

In 1990 Alan Grafen published a pair of papers in *Journal of Theoretical Biology* (Grafen 1990a, 1990c; more than

70 pages total) which demonstrated both that honest signals can exist and that under some conditions cost can ensure their stability. Unfortunately, few could read these highly mathematical papers. The paper that blew open the doors of costly signaling for biologists was John Maynard Smith's analysis of the Sir Philip Sidney game.[58] In his usual style Maynard Smith boiled the problem down to a simple model involving only two pages of high-school algebra.

We'll examine the Sir Philip Sidney game because it nicely illustrates some general properties of costly signaling models. It is based on a tale British children sometimes learn in school. Sir Philip Sidney was an Elizabethan courtier, poet, diplomat, and soldier who was mortally wounded in a skirmish with the Spanish at Zutphen in 1586. Before he died Sidney gave his canteen to a wounded soldier nearby, saying "Thy need is greater than mine," an action Maynard Smith describes as "an unusual example of altruism by a member of the British upper classes." As we will see, Maynard Smith's model really doesn't fit this story very well. However, the model does apply to important biological situations like parental care, competition between siblings, and communal breeding.

Assume that there are two players, a potential donor D and a possible beneficiary B. The donor has a resource, in Sir Philip's case a water bottle, which he can give to the beneficiary. Doing so reduces the donor's survival. If the donor keeps the bottle, he has fitness 1. If he gives it to B, he has fitness $1 - d$. Thus the cost to the donor of transferring the resource is d. (Setting the baseline fitness to 1 will simplify the expressions below. We can do this without any loss in generality because it is in effect setting the units of measurement of fitness.)

The beneficiary benefits from receiving the resource (the water bottle), but how much he benefits depends on his condition. If he gets the resource, he has the baseline fitness 1 no matter what his condition. If B does not get the resource, his fitness depends upon his state. If he is severely wounded,

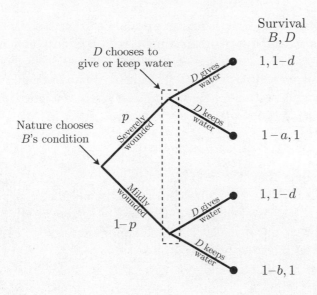

Figure 5.1: Structure of and payoffs in the Sir Philip Sidney game. Variables are defined in the text.

his fitness will be $1 - a$. But if he is only mildly wounded, it will be $1 - b$. Assume that B is severely wounded a fraction p of the time. B knows his true state, but the donor does not know B's state. Figure 5.1 summarizes these effects.

For it to ever be in the donor's interest to transfer the resource, D has to have some interest in B's survival. Maynard Smith assumes that D and B are related by a coefficient of relatedness r. For example, it could be that B is a parent and D is its child. (This is the problem with story. Its not clear why Sir Philip should be related to one of his men-at-arms. Perhaps his dad had been busy downstairs as well as up.) Let's map out the spectrum of conflict and common interest. Consider when the donor would want to transfer the resource, assuming it knew B's condition with certainty. When B is mildly wounded, D will want to transfer the resource provided

$$r \times \text{(change in fitness of } B) > \text{(change in fitness of } D)$$
$$rb > d.$$

We assume that the cost to the donor is greater than the cost to a mildly wounded beneficiary, $d > b$, which means that this condition is never satisfied ($rb < d$). D would not give the water bottle to B if he knew for sure that B was only mildly wounded. When B is severely wounded D will want the transfer when

$$ra > d.$$

We assume that $d < a$. This means that D would favor the transfer if he knew for sure that B was in bad shape, provided the two are sufficiently related, $r > d/a$. These assumptions ($b < d < a$) mean that it is in the donor's interest to discriminate based on the beneficiary's state.

The beneficiary will not have the same view of things. When B is mildly wounded, he favors the transfer when

$$rd < b.$$

If r is too low ($r < b/d$), B will want the resource even when he is only mildly wounded. When he is severely wounded, B will want the resource when

$$rd < a.$$

This condition is always satisfied, since $a > d$, so B always wants the resource when he is severely wounded. This makes sense, since he needs it more than the donor, and he is more closely related to himself.

Thus there are three regions of relatedness leading to different amounts of conflict over the transfer (Figure 5.2). On the left in Figure 5.2, when D and B are not closely related ($r < d/a$), D never wants to donate and B always wants the water bottle. In this region, signaling can never be honest, and D should never believe any signal from B. On the right, when $r > b/d$, D and B are closely enough related that they both agree that B should only get the resource if he is severely wounded. Thus signals in this region should always be honest, since there is no conflict of interest between the

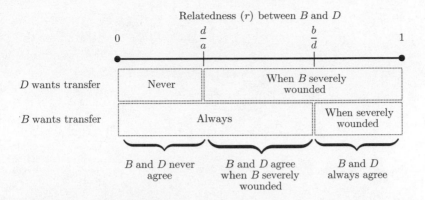

Figure 5.2: Common and conflicting interests, in the Sir Philip Sidney game, depending upon the relatedness (r) between the donor (D) and the beneficiary (B).

two parties. These are Tinbergen's low-cost, honest signals. In the middle, however, D only wants to transfer the resource when B is severely wounded, but B would always benefit from the transfer, regardless of the severity of his injury. In this region there is a conflict of interest.

This model applies to many examples of parental care. Parents often control resources that they could either keep for their own benefit or give to their offspring. However, only the offspring knows for certain how much it really needs the resource. This leads to a conflict of interest between parents and offspring. Think of chicks begging in a nest. Loud baby birds announce their presence to predators, so why should they chirp at all? One answer is to say "I'm hungry." But then wouldn't the baby bird always say "I'm hungry"? And if so, why would the parent ever listen? Any of you with children of your own or with younger brothers or sisters will recognize the same dilemma in human child care.

To model such "begging," we allow the beneficiary to signal his state. After nature chooses B's condition, he can signal, at a personal cost c, that he needs the resource. He can also remain silent and incur no cost. The donor then

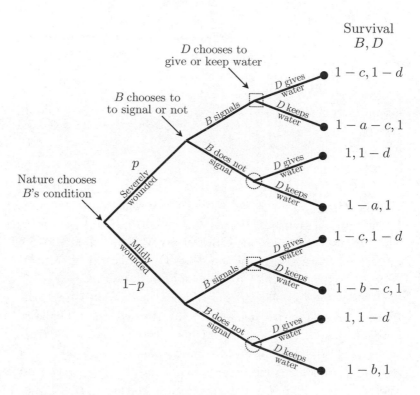

Figure 5.3: Structure of and payoffs in the Sir Philip Sidney game when signaling reduces B's survival by an amount c. Dotted squares and circles represent separate information sets.

has a chance to respond by giving or keeping the resource. The consequences of adding costly signaling to the model are shown in Figure 5.3. Given that B will always benefit from getting the resource, when will natural selection favor honest signaling of state? The answer is that honest signaling can be an ESS provided that (1) the signals are sufficiently costly or (2) there is sufficient common (genetic, in this case) interest between D and B. In the remainder of this section, we'll derive these results.

Consider a strategy called Responder (R). When Responders are beneficiaries, they only signal when severely wounded. When they are donors, they only transfer if the beneficiary signals. First, consider when R can resist invasion by rare mutants that ignore the signal when they are the donor. There are two possibilities: one mutant always gives (AG) the resource and the other never gives (NG) it. These mutants continue to signal only when needy when they are beneficiaries. We will ignore the donor who always gives because this strategy is assumed to be a bad idea already. Instead we'll only ask when a donor who never gives (NG) can invade one who responds to signals. We'll write the payoff to a donor with strategy X playing with a beneficiary with strategy Y as $V_D(X|Y)$. There are two possible payoffs:

$$V_D(R|R) = (1-p)(1) + p(1-d)$$
$$V_D(NG|R) = (1-p)(1) + p(1).$$

Similarly, there are the payoffs to the beneficiaries using strategy X when playing a donor with strategy Y, $V_B(X|Y)$:

$$V_B(R|R) = (1-p)(1-b) + p(1-c)$$
$$V_B(R|NG) = (1-p)(1-b) + p(1-a-c).$$

R can resist invasion by NG if

$$\underbrace{V_D(R|R) + rV_B(R|R)}_{\text{inclusive fitness of } D \text{ who responds}} > \underbrace{V_D(NG|R) + rV_B(R|NG)}_{\text{inclusive fitness of } D \text{ who never gives}}$$
$$1 - dp + r(1 - (1-p)b - pc) > 1 + r(1 - (1-p)b - p(a+c)).$$

Solving for r gives the condition for R to be stable:

$$r > \frac{d}{a}.$$

When r is less than this threshold value there isn't enough common interest, and it will pay to always keep the resource, regardless of the condition of the beneficiary.

Now for the interesting part: when can liars invade? Here we consider whether a population in which R is common can resist rare mutants that modify the behavior of beneficiaries. Once again there are two possibilities: never signal (NS) or always signal (AS). We are interested in the beneficiaries who alway signals (AS)—these are the liars. As before, we can define four payoffs:

$$V_D(R|R) = (1 - p)(1) + p(1 - d)$$
$$V_D(R|AS) = (1 - p)(1 - d) + p(1 - d)$$
$$V_B(R|R) = (1 - p)(1 - b) + p(1 - c)$$
$$V_B(AS|R) = (1 - p)(1 - c) + p(1 - c).$$

Again, use the inclusive fitness rule to figure out when R can resist invasion by AS:

$$\underbrace{rV_D(R|R) + V_B(R|R)}_{\text{inclusive fitness of honest } B} > \underbrace{rV_D(R|AS) + V_B(AS|R)}_{\text{inclusive fitness of lying } B}.$$

Note now that we weight the fitness of the donor by r, instead of the fitness of the beneficiary. We do this because the genes that control signaling are expressed in B. Substituting the fitnesses gives

$$r\underbrace{(1 - dp)}_{V_D(R|R)} + \underbrace{1 - (1 - p)b - pc}_{V_B(R|R)} > r\underbrace{(1 - d)}_{V_D(R|AS)} + \underbrace{1 - c}_{V_B(AS|R)}.$$

This expression simplifies to

$$r > \frac{b - c}{d}.$$

Thus as long as $c > b - rd$, honest signaling can resist invasion by liars who signal when they are in good condition. This condition makes sense because $b - rd$ is the inclusive fitness increase that the beneficiary gets by receiving the resource when he is in good condition. Thus as long as the signal costs at least this much, there is no motive to lie. And, as we saw above, if there is no motive to lie, donors will respond honestly to the signal.

So the model tells us that some signals can be honest as long as they are costly. But that's not the whole story; it also tells us two interesting facts. First, notice that none of the conditions we derived for the stability of honest signaling depended on p, the probability of being in need. In other words, it doesn't matter whether beneficiaries are in need 50 percent the time or 0.1 percent of the time. This result seems weird at first because the need for signals most definitely does depend on p. If the other guy is in need half the time, "inquiring minds want to know," but if he is in need only 0.1 percent of the time, there's not much benefit to discriminating—"just say no." But when you think about it a little bit this result makes sense. The role of the costly signal is to deter liars. It doesn't matter how many liars there are out there. Even if there is only one, he will be believable only if he has to pay enough to preclude lies.

The second interesting fact is that there is always a non-signaling equilibrium as well (a fact you will prove when you solve Problem 1 at the end of this chapter), and there is a plausible range of parameter combinations at which the nonsignaling equilibrium yields higher payoffs for both the beneficiary and the donor (a fact you will prove when you solve Problem 2 at the end of this chapter). As emphasized by Carl Bergstrom and Michael Lachmann,[59] these two facts suggest that we need to be careful when applying costly signaling theory to understand animal signals. To see why, it will be useful to consider a model in which both donors and signalers can take on many different states.

5.1.2   The continuous Sir Philip Sidney game

The Sir Philip Sydney game provides a very simple illustration of the basic principles of costly signaling theory. However, it is a bit too simple to expose all of the interesting principles. So, now we turn to a model in which both the donor and the beneficiary can take on a continuum of states.[60] We will see that, as in the discrete version of the game, there is a costly signaling equilibrium at which beneficiaries truthfully and accurately signal their state. However, we will also see that there are many other equilibria at which individuals in a range of states give the same signal and individuals in other ranges give other signals. Beneficiaries can distinguish individuals giving different signals, but not among individuals giving the same signal. The first type of equilibria are called *separating* equilibria and the second type are called *pooling* equilibria. Pooling equilibria may be very important in nature because they can be much less costly than separating equilibria. In fact, under the right circumstances, they can even support cost-free, honest signals.

Once again there are two players, a potential donor D and a potential beneficiary B who are related by r. D has a resource that she can keep or transfer to B. Whichever one ends up with the resource has fitness 1. If she doesn't get the resource, B has fitness x and D has fitness y; x and y are values in the interval $[0, 1]$. We will refer to an individual's fitness without the resource as its "condition." Each individual knows its own condition but only knows that the other individual's is drawn from a uniform probability distribution. This means that all values of x and y from 0 to 1 are equally likely. We assume uniform distributions to keep the math from getting too difficult, but Bergstrom and Lachmann[61] showed that you get the same qualitative results for any probability distribution.

To begin, let's figure out where the conflict between donor and beneficiary lies. First, imagine that the donor knows the condition of the beneficiary. If she transfers the resource, the donor's inclusive fitness is $V_D(T) = y + r(1)$. If she instead keeps the resource, her inclusive fitness is $V_D(K) = 1 + rx$. Thus the donor transfers if $V_D(T) > V_D(K)$:

$$y > 1 - r + rx.$$

If the donor is in poor enough condition, y is low enough, then he will not prefer a transfer. However, as r increases, interest in the beneficiary's condition increases, and the value of x has a higher impact on the outcome. If x is small and r is large, the donor might favor the transfer over a wide range of y.

Of course, the donor does not actually know x. So next consider what happens if the donor has to guess. If she doesn't transfer the resource, her inclusive fitness doesn't depend on the fitness of the beneficiary and thus is still $y + r$. However, if she does transfer, she (or rather selection) knows the average effect because the best guess of the beneficiary's condition will be the expected value of x—the average condition of a large number of beneficiaries. Thus the expected inclusive fitness of keeping the resource will be

$$\mathrm{E}(1 + rx) = 1 + r\overline{x},$$

where $\overline{x} = 1/2$ is the average condition of beneficiaries. A donor will transfer only if $y > 1 - r + r\overline{x} = 1 - \frac{1}{2}r$. Call this threshold value $y_{\min}$. As is shown in Figure 5.4, this means that honest signals only make a difference for some combinations of x and y. If the donor has condition $y < y_{\min}$ and the beneficiary has $x < \overline{x}$, knowledge of the beneficiary's actual state might induce the donor to transfer the resource. In contrast if the donor's state is $y > y_{\min}$ and the beneficiary's is $x > \overline{x}$, perfect knowledge would lead the donor to keep the resource where he would have transferred it in ignorance. In all other cases, signals cannot affect the outcome because either the conflict of interest ($y < y_{\min}$, $x > \overline{x}$) or degree of coordinated interest ($y > y_{\min}$, $x < \overline{x}$) is too great.

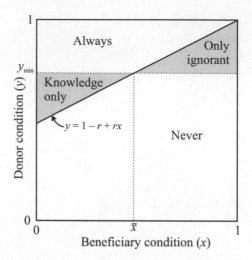

Figure 5.4: Each point represents a combination of donor and beneficiary fitness when each lacks the resource. The different regions show the conditions in which the donor would favor the transfer. Donors who know the fitness of the beneficiary will transfer if $y > 1 - r + rx$, i.e. above the sloping line. Only donors whose own fitness is greater than $y_{\min} = 1 - r + r\overline{x}$ will donate when they are ignorant of the beneficiary's fitness. Thus, signals only make a difference in the shaded areas. The figure is plotted for $r = 1/2$. Lower levels of relatedness decrease the regions in which signaling makes a difference because relatedness determines the slope of the line.

The separating equilibrium

Now we are ready to determine the signal cost function $c(x)$ that allows an individual with fitness x to honestly and precisely signal her condition without liars being able to invade— the separating equilibrium. This is analogous to deriving the condition for Responder to resist invasion by Always Signal in the discrete model of the last section.

Focus on an individual with a different need $x' = x + \delta$ that is a little bigger than x. An honest signal will have cost $c(x')$ and a dishonest signal will have a higher cost $c(x)$. How much more costly must the lie be for honesty to pay? In Box 5.1 we show that the rate of cost decrease (as a function of x) that is just enough make an honest signaler better off than a dishonest one is given by

$$-r(1-x)(1-r^2) = \frac{dc}{dx}. \tag{5.1}$$

This condition says that to deter dishonesty, as the condition of the beneficiary improves, the cost of signaling has to decrease at the rate given on the left-hand side of Equation 5.1.

This is a first-order differential equation, and to solve it we need to know the value of $c(x)$ for some value of x—this is called a "boundary condition." Here the obvious condition is $c(1) = c_{\min}$. Beneficiaries who are in the best possible condition (i.e., $x = 1$) are never going to get the resource, so they make the cheapest possible signal. Because not signaling can be a signal, it may often be that $c_{\min} = 0$. Once we know the value at $x = 1$, we can integrate backwards, calculating each value of $c(x)$ from the one before it. In other words, we just computed the rate of change in cost as a function of x. This function is a derivative. If we integrate this function with respect to x, we get the total cost function back:

$$c(x) = \int \frac{dc}{dx}dx = \int r(x-1)(1-r^2)dx$$
$$= \frac{1}{2}r(1-r^2)(x-1)^2 + k,$$

where k is an unknown constant whose value is determined by setting $c(1) = c_{\min}$. Thus, $k = c_{\min}$.

When individuals in condition x signal with cost $c(x)$, all signals are honest and perfectly reveal the player's condition. Donors who receive a signal with cost $c(x)$ will transfer the resource if $y > 1 - r + rx$. Thus this analysis predicts that

Box 5.1 Determining the cost function $c(x)$

For notational convenience, let $a = 1 - r + rx$ and $b = 1 - r + rx'$. An honest individual with condition $x' = x + \delta$, where $\delta > 0$, will have higher inclusive fitness giving the honest signal $c(x')$ rather than the dishonest signal $c(x) > c(x')$ if

$$\Pr(y < b)(x' + r) + \Pr(y > b)\{1 + r\,\mathrm{E}(y|y > b)\} - c(x')$$
$$> \Pr(y < a)(x' + r) + \Pr(y > a)\{1 + r\,\mathrm{E}(y|y > a)\} - c(x).$$

The probabilities take the average over donors in different conditions. The first term on each side of the inequality is the average for donors who do not transfer—their own condition y is less than the threshold that makes it in their interest—and the second term is that for those who do transfer. The liar gets the resource from donors whose condition is in the range $[a, 1]$ and the honest signaler gets the resource from donors whose condition is in the smaller range $[b, 1]$. Luckily, only what happens in the range $[a, b]$ determines which strategy has higher inclusive fitness. It is in this range that the liar gets more benefit but pays greater cost. In all other ranges, the two are equal. Using this fact, we rewrite the above condition as

$$\Pr(a < y < b)(x' + r) - c(x')$$
$$> \Pr(a < y < b)\{1 + r\,\mathrm{E}(y|a < y < b)\} - c(x)$$
$$\Pr(a < y < b)\{1 + r\,\mathrm{E}(y|a < y < b) - x' - r\} < c(x) - c(x').$$

The left-hand side of the inequality represents the increase in inclusive fitness produced by the lie, and the right-hand side represents the increased cost experienced by a liar who signals she is in worse condition than she really is. Since y is a uniformly distributed random value between 0 and 1, $\Pr(a < y < b) = \frac{b-a}{1-0} = r\delta$. Now we compute the expected value of y in this range, $\mathrm{E}(y|a < y < b)$, which is merely the average of a and b, and simplify (remembering to cancel all terms of order δ^2):

$$r\delta\left(1 + r\frac{a+b}{2} - x - \delta - r\right) < c(x) - c(x + \delta)$$
$$r\delta\big(1 + r(1 - r + rx) - x - r\big) < c(x) - c(x) - \delta\frac{dc}{dx}.$$

Simplifying the above yields Expression 5.1 in the main text.

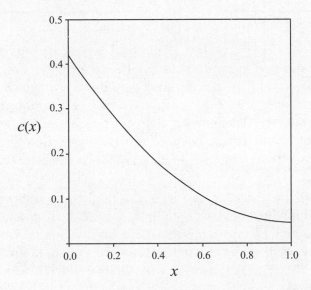

Figure 5.5: Plots the form $c(x)$ at the separating equilibrium for $r = 1/2$ and $c_{min} = 0.05$. These are the least costly signals for each value of x that allow accurate communication. Even so, notice that for individuals in poor condition the cost of signaling consumes almost half of the value of getting the resource. Consider a beneficiary in condition $x = 0.2$. The expected gain from signaling is $1 - 0.2 - c(0.2) \approx 0.5$. The resource itself is worth 0.8 to this individual. The expected benefit from making a costly signal can be less than the value of the resource because many individuals will signal but only some will be rewarded.

signals will be graded in cost. More and more needy individuals will emit more and more costly signals, and donors will make fine discriminations based on the cost of the signal. As can be seen in Figure 5.5, the costs of signaling can amount to a significant fraction of the benefit of transferring the resource, and as a result, even though costly signals allow completely accurate communication, they may not bring much benefit. This means that selection can favor less informative but cheaper communication, the subject to which we now turn.

Pooling equilibria

We have already seen one example of a pooling equilibrium. In the discrete version of the Sir Philip Sidney game, there were two equilibria: a separating equilibrium, in which costly signals ensure signals of need, and a pooling equilibrium, in which nobody signaled. In the continuous game, there are also both separating and pooling equilibria. However, now there is a vast range of pooling equilibria rather than just one.

Here we consider the simplest pooling equilibrium, in which individuals whose conditions exceed a threshold value a (i.e., $x > a$) give one signal at a cost c_2 and individuals whose condition values are less than the threshold give a signal at cost c_1. We want to find the the difference between these costs, $c_1 - c_2$, so that only individuals with condition $x < a$ can afford to signal. To do this, notice that individuals in the poorest condition among the nonsignalers will be most tempted to lie. So, the cost of giving the signal, c, must be high enough to deter individuals with $x = a$ not to signal. This requires

$$c_1 - c_2 = f\big((1 - a) - r(1 - \overline{y}_t)\big),$$

where f is the fraction of donors who transfer when they receive a signal but don't transfer if they don't receive a signal, and $\overline{y}_t$ is the average fitness of those donors. Thus $c_1 - c_2$ is the inclusive fitness benefit that would accrue to a signaler with condition a. The term $1 - a$ is her increase in fitness from getting the resource, and $1 - \overline{y}_t$ is the decrease in the donor's fitness. Signalers only get this benefit in a fraction f of the times they signal. Thus, it does not pay for individuals with condition better than a to signal, but it does pay those whose condition is less than a.

First we calculate f. Because the donors can't distinguish among the individuals who signal, they respond to the average condition of those who signal, $\overline{x}_1 = \frac{0+a}{2} = a/2$. Assuming the signal is accurate, donors' inclusive fitness will increase if

$$y_1 > 1 - r + r\overline{x}_1 = 1 - r + r\left(\frac{a}{2}\right).$$

Similarly, the average condition of those who do not signal is $\overline{x}_2 = (1 + a)/2$, and thus donors will transfer to individuals who do not signal if

$$y_2 > 1 - r + r\overline{x}_2 = 1 - r + r\left(\frac{1+a}{2}\right).$$

Since y is uniformly distributed, $1 - y_1$ is the fraction of donors who give when they receive a signal and $1 - y_2$ is the fraction who give when they do not receive a signal. This means that by giving the signal, beneficiaries increase the fraction of donors who will transfer to them by

$$f = (1 - y_1) - (1 - y_2) = y_2 - y_1 = \frac{r}{2}.$$

The average condition of those who transfer the resource is

$$\overline{y}_t = \frac{1}{2}(y_1 + y_2) = 1 - \frac{3}{4}r + \frac{a}{2}r.$$

Substituting the expressions for f and $\overline{y}_t$, we find that the equilibrium cost of signaling is

$$c_1 - c_2 = \frac{r}{2}\left(1 - a + r^2\left(\frac{a}{2} - \frac{3}{4}\right)\right).$$

This cost difference guarantees that everyone whose condition is less than the threshold can profitably signal this fact, and those whose condition is above the threshold cannot. Donors will respond according to the average condition in each pool and their own condition.

There is nothing biologically implausible about such pooling equilibria. Unlike the separating equilibria, cost in this pooling equilibrium does depend on the distribution of condition among signalers. And because not all types (all values of x) have to be distinguished, the cost of signaling can be much lower. In fact, as we will see in the next section, it can be zero.

5.2 Cheap, honest signals

Some people seem to have interpreted the success of Zahavi's idea as "signals can be honest *only* if they are costly." Maynard Smith's model makes it clear that this isn't true: if the parties share sufficient common interest, there is no motivation to lie. A greater number of people have acquired the slightly less wrong belief that Zahavi's idea indicates that, when interests conflict, signals can only be honest if they are costly to produce.

It turns out, however, that honest, cheap signals can be evolutionary stable even when interests conflict. In this section, we'll take a look at two different reasons. First, we will see that there are always many pooling equilibria in the continuous Sir Philip Sidney game that allow zero-cost signaling. In effect, the right pooling converts a situation with sharp conflicts of interest into one with much attenuated conflicts. Second, we consider a very simple model in which there are substantial conflicts of interest, but repeated interaction allows honest signaling. In Problem 3 at the end of this chapter, you will show that honest signaling can be stable when there are conflicts of interest, as long as there is an even stronger interest in coordination.

5.2.1 Cheap talk in the pool

In the last section, we saw that there is an equilibrium in the continuous Sir Philip Sidney game in which beneficiaries are divided into two groups. Those with condition less than a threshold a give one signal at a cost c_1 and those above the threshold give a different signal with cost c_2. It turns out that the right choice of threshold allows $c_1 = c_2$—no difference in the cost of the two signals! Strange but true. Just set

$$\frac{r}{2}\left(1 - a_0 + r^2\left(\frac{a_0}{2} - \frac{3}{4}\right)\right) = 0$$

and solve for a_0. This yields

$$a_0 = \frac{1 - \frac{3r}{4}}{1 - \frac{r^2}{2}}.$$

If x is greater than a_0, individuals give one signal. If x is less, they give a different one, but there is no cost difference, and thus no penalty for lying. As long as $r > 0$, $a_0 < 1$, and there is a pooling equilibrium that allows costless, honest signaling. What's more, Bergstrom and Lachmann[62] showed that there are also pooling equilibria that have more than two partitions and still have no cost difference in the signals given by individuals belonging to different partitions.

What the heck is going on here? If it doesn't cost them anything, why don't individuals in the less needy pool signal that they are needy? The answer is that they *would* increase the chance they get the resource, but in doing so they would, on average, reduce the fitness of the donors enough that their inclusive fitness decreases. Figure 5.6 illustrates why. Once again we plot the beneficiary's condition on the x-axis and the donor's condition on the y-axis. As before, above the upper line, $y = 1 - r + rx$, both donor and recipient increase their inclusive fitness by making the transfer. Below the second line, $y = 1 - 1/r + x/r$, the beneficiary's inclusive fitness is higher if the transfer does not take place. (If the donor transfers, the beneficiary's inclusive fitness is $ry + 1$; if she doesn't, the beneficiary's inclusive fitness is $r + x$. Thus the beneficiary prefers that the donor keep the resource if $y > 1 - 1/r + x/r$.)

Now consider a beneficiary whose condition is just a tiny bit above the threshold a_0 that divides the two beneficiaries into good- and bad-condition classes. This is the individual who would most benefit from signaling. As a member of the good condition pool, she only gets transfers from donors whose condition exceeds $\bar{y}_2$. Ideally she would like to signal her true condition so she could get transfers from donors whose condition was greater than $y^\star$. However, there is no signal that will be understood that allows her to do this. Her

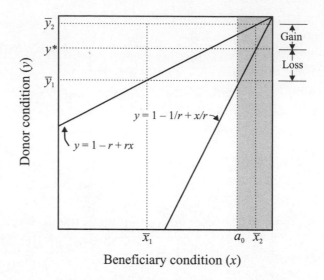

Figure 5.6: A beneficiary whose condition is just a bit above the threshold a_0 signals that she is in good condition, which means that she gets transfers from donors whose condition exceeds $\bar{y}_2$. If she signals that she is in poor condition, she gets contributions from more donors, those whose condition exceeds $\bar{y}_1$. Transfers from donors whose condition lies above $y^\star = 1 - 1/r + a_0/r$ increase her inclusive fitness. Transfers from donors below $y^\star$ decrease it. The threshold between the pools is such that the losses always exceed the gains, and thus selection favors honest, zero-cost signals.

only option is to signal that she is in the low-condition pool, which means that she will get transfers from donors whose condition exceeds $\bar{y}_1$. Transfers from donors between $y^\star$ and $\bar{y}_1$ lead to a loss in the beneficiary's inclusive fitness that is greater than the gains she would achieve from transfers from donors between $y^\star$ and $\bar{y}_2$. It's like the discrete game in which cheap talk was favored as long as individuals were sufficiently closely related. But now, because the threshold between the pools can evolve, there is always a partition that allows cheap, honest signals.

It's not clear whether this kind of cheap talk is important in the real world. Studies of begging by nestlings seem to suggest that begging is not especially costly energetically, consistent with this model. However, begging takes time and attracts predators, costs that are very difficult to measure. There are also a number of theoretical worries. Brilot and Johnstone[63] pointed out that the existence of zero-cost pooling equilibria is sensitive to the existence of donors with no need for the resource. Also, the costs of separating equilibria are reduced when more than one sib competes, so that the relative benefit of pooling is also decreased.[64] Finally, Maynard Smith and Harper argued that donors might tend to distinguish between different signal intensities within each class as a side effect of the design of their sensory systems. If so, selection will favor beneficiaries who expend cost to distinguish themselves from others, and this will destabilize the pooling equilibrium.

5.2.2 Repeat play and cheap signaling

So far all of the signaling models have assumed one-shot interaction. Individuals meet, someone signals, and they never meet again. But in many interesting interactions individuals meet repeatedly, and when they do, selection can favor low-cost communication. The reason is obvious—someone who lies won't be believed in the future, and as a result will forgo the benefits that result from communication. In other words, never cry wolf (unless the wolf is really there).

Here we consider a model of signaling based on interactions between individuals within macaque and baboon groups.[65] In these monkeys, one individual often approaches a second with the intent of engaging in a mutually beneficial interaction. The problem is that the same individual sometimes approaches with the intent of attacking the other monkey. To solve this problem, macaques and baboons give quiet signals that reliably indicate peaceful intent. Why aren't these cheap, quiet signals destabilized by lies?

To model this situation assume that individuals interact in pairs, and one individual (the actor) approaches the other (the recipient) and may produce a signal. The actor can have one of two intentions: peaceful or hostile. The actor is peaceful p of the time and hostile $1 - p$ of the time. Assume that intentions are unknown to the recipient. We will assume here that signal production is costless.

After being approached and receiving a signal, the recipient has the opportunity to stay and interact with the actor or flee. There are two sets of payoffs, one when the recipient decides to stay and one when she decides to flee. When she stays, if the actor is peaceful (p of the time), they interact peacefully, and the actor gets I (benefit of interaction) and the recipient i. If the actor is hostile, they fight, and the actor receives E and the recipient $-e$. These payoffs are meant to represent a situation in which a dominant animal is approaching a subordinate animal. Thus the actor wins the fight and receives the positive payoff, while the recipient suffers. The recipient can flee before the actor has a chance to act aggressively, however. When she flees, the actor receives D (the benefit of displacing the recipient) and the recipient receives $-d$ (the cost of being displaced).

Without repeat interaction. The actor can pursue one of three strategies: Truthful (TF), which signals when of peaceful intent, Never Signal (NS), and Always Signal (AS). Recipients similarly have three strategies: Always Interact (AI), Always Flee (AF), and Believe (BV), which interacts when the actor signals and flees when the actor does not signal.

Consider when TF is an ESS, given that recipients are always BV. The fitnesses of the actor strategies against BV are, respective.

$$V(TF|BV) = pI + (1 - p)D$$
$$V(AS|BV) = pI + (1 - p)E$$
$$V(NS|BV) = pD + (1 - p)D.$$

TF is an ESS when $V(TF|BV)$ is greater than both $V(AS|BV)$ and $V(NS|BV)$. For TF to resist AS we must have

$$pI + (1-p)D > pI + (1-p)E$$
$$D > E.$$

This means that for TF to be an ESS against AS, the benefit of displacing must exceed the benefit of aggression. For TF to be an ESS against NS we must have

$$pI + (1-p)D > pD + (1-p)D$$
$$I > D.$$

Thus for TF to resist invasion by NS, the benefits of interaction must exceed the benefits of displacing the recipient. Taken together, these two conditions say that honest signaling is an ESS only if the actor has nothing to gain from lying. If the benefit of a fight relative to displacement is ever positive $(E - D > 0)$, then honest signaling cannot be an ESS.

The recipient will believe the signal when the payoff to BV is greater than the payoffs to AI and AF. The three recipient payoffs are

$$V(BV|TF) = pi - (1-p)d$$
$$V(AI|TF) = pi - (1-p)e$$
$$V(AF|TF) = -pd - (1-p)d.$$

These payoffs then imply that BV is an ESS when

$$i > -d > -e.$$

This is the same order of payoffs that make TF an ESS: $I > D > E$. The above model is similar to the Sir Philip Sidney game, but without the costs of signaling. For honest signaling to be an ESS, both parties must order payoffs in the same way. That is, they must have the same interests. Otherwise, costly signaling would be necessary to stabilize the signals.

With repeat interaction. However, instead of allowing for costly signals, let's add repeat interaction to the model. Assume that pairs of individuals are formed at random and have a chance w of continuing on to another interaction, just as we modeled the iterated prisoner's dilemma in Chapter 4. Cheap talk allowed the boy who cried wolf to get a good laugh, but when the wolf really arrived, he lost all his sheep. Aesop's moral: liars will not be believed. To formalize this idea, substitute a new strategy, Conditional Believer (CB), for Believer (BV). CB individuals believe until they are deceived. After a deception, they always flee the actor.

Consider first when CB is an ESS when TF is common. The payoffs are

$$V(CB|TF) = pi - (1-p)d + wV(CB|TF)$$
$$= \frac{pi - (1-p)d}{1-w},$$

$$V(AI|TF) = pi - (1-p)e + wV(AI|TF)$$
$$= \frac{pi - (1-p)e}{1-w},$$

$$V(AF|TF) = -d + wV(AF|TF)$$
$$= \frac{-d}{1-w}.$$

Thus CB is an ESS provided

$$i > -d > -e.$$

This is just as before. If the recipient would rather interact than be displaced, and rather be displaced than get beat up, responding to signals is stable.

The real effect of repeat interaction lies instead in the dynamics of signaler strategies. When will TF be an ESS?

The payoffs are now

$$V(TF|CB) = \frac{pI + (1-p)D}{1-w}$$

$$V(AS|CB) = p\{I + wV(AS|CB)\}$$
$$+ (1-p)\{E + wV(AS|AF)\} \qquad (5.2)$$

$$= \frac{pI + (1-p)E + (1-p)\dfrac{wD}{1-w}}{1-wp} \qquad (5.3)$$

$$V(NS|CB) = \frac{D}{1-w}.$$

Equation 5.2 deserves some explanation. When the actor has good intentions (p of the time), she signals, and CB remains and interacts, yielding I for the actor. If they go on to interact another round (w of the time), the expected payoff is the same because the actor has not yet lied. However, when the actor has darker intentions ($1-p$ of the time), she signals, and CB remains, yielding E for the actor. However, if they go on to interact again, CB will never again respond to signals, so the expected payoff is as if the recipient were playing Always Flee. Always Signal against Always Flee gets D every round, for as long as interactions last. Combining all this, and using the techiques introduced in Chapter 4, we get Equation 5.3.

Using the above payoffs, we find that TF is an ESS against AS when

$$\frac{pI + (1-p)D}{1-w} > \frac{pI + (1-p)E + (1-p)\dfrac{wD}{1-w}}{1-wp}$$

$$\frac{w}{1-w}p(I-D) > E - D. \qquad (5.4)$$

The left side of Expression 5.4 is the long-term benefit of honesty. This is because if the actor lies, she will get D in every turn after the first, so $I - D$ is the difference in payoff from being honest. The right side is the average gain from lying in the first turn. When the long-term benefits of honesty exceed the benefits of lying in the first turn, honesty is an ESS. This may seem like common sense, but recall that our previous models argued that honesty could not be an ESS unless the actor and recipient had the same ordering of payoffs. This need not be the case here. It is fine to assume that for the recipient it is better to flee $(-d)$ than to be beaten up $(-e)$, yet the condition above allows TF to be an ESS even if $E > D$, meaning that the actor would benefit more from beating up the recipient.

This result shows that honest signals can sometimes exist even if the parties have insufficient common interest and lying is not immediately costly. In long-lived social species like many birds and primates, we should probably be careful about concluding that cheap, honest communication should be a feature of short-term coordinated interest only. Provided that individuals remember past interactions and recognize other individuals, cheap, honest signals may be possible.

5.2.3   Costly signals need not indicate conflicts of interest

Interestingly, Johnstone[66] also showed that costly signals may evolve in contexts of strong common interest, not just when interests conflict. Cost in his model is not needed to keep signals honest, but rather to make them reliable. If there is noise in production or reception of the signal, selection may favor repetition and reinforcement to ensure it is understood. This casts doubt on the intuition that costly signals will be observed only where interests conflict and in order to ensure honesty. This means we cannot infer from the existence of costly or time-consuming signaling that animals have different interests.

If you think about this a moment, it must be true. If some matter is important enough to the common interest of two (or more) individuals, then they should spend as much time as it takes to make sure they understand one another and do the right thing. Thus, speaking up or repeating oneself can also be adaptations to common interest, not just solutions to a conflict of interest.

5.3   Signaling and altruism

There is one possible solution to the problem of altruism we studied in Chapters 3 and 4 which seems to reoccur again and again to people in many different disciplines. If altruists could signal they were altruists, then altruists could direct help to those who give the signal. Suppose, for example, in one version of the story, that altruists wore a "green beard."[67] Altruists would then adopt the rule "help only individuals who wear green beards." The fitness of an altruist in this model is

$$W(A) = p(b - c) + (1 - p)(0) + w_0,$$

where p is the frequency of altruists. The fitness of a non-altruist is

$$W(N) = p(0) + (1 - p)(0) + w_0.$$

Thus altruists have higher fitness than nonaltruists at all values of p, and altruists will invade and be an ESS. This wouldn't be kin selection and wouldn't rely upon repeat interaction. It should work even in very simple organisms who only interact once. If this argument is right, why isn't altruism everywhere in nature?

The green beard idea has a fatal weakness. It is very unlikely that a single gene could both produce the signal (the green beard) and code for the behavior (be nice to people with green beards). Even worse, if there was a such a gene, it would have to be true that no other gene anywhere in the genome could modify the phenotype so that nonaltruists had a green beard. Our usual game-theoretic assumption that a strategy can be complex and adequately represented by a single gene breaks down in this case. In this section, we'll see how serious this weakness is. Along the way, we'll learn how to model evolution at two genetic loci. This will be very useful because sometimes we are interested in gene–gene (or gene–culture) interactions, and without some mathematical machinery for concisely expressing such a system, analytic results would be hard to come by.

Imagine a haploid organism with two loci: one controls the development of a signal and another controls behavior when the signal is received. There are two alleles at each locus. At the first, which controls behavior, there are altruists (A) and nonaltruists (N). At the second, which controls the signal, there are signalers (G) and nonsignalers (N). This implies four different genotypes:

Genotype	Phenotype
NN	Nonaltruist, no signal
NG	Nonaltruist, signals
AN	Altruist, no signal
AG	Altruist, signals

Define the frequency and fitness of each genotype as follows:

Genotype	Frequency	Fitness
NN	x_{00}	W_{00}
NG	x_{01}	W_{01}
AN	x_{10}	W_{10}
AG	x_{11}	W_{11}

Let the frequency of altruist alleles in the population be

$$p = x_{11} + x_{10}$$

and let the frequency of green beards (G) be

$$q = x_{11} + x_{01}.$$

The fitness of each genotype is then

$$W_{00} = w_0$$
$$W_{01} = pb + w_0$$
$$W_{10} = q(-c) + w_0$$
$$W_{11} = pb - qc + w_0.$$

Box 5.2 Deriving the covariance between A and G
We can compute the required expectation with a table:

Genotype	Freq	p_i	q_i	$p_i q_i$
NN	x_{00}	0	0	0
NG	x_{01}	0	1	0
AN	x_{10}	1	0	0
AG	x_{11}	1	1	1

We get the expectation of $p_i q_i$ by adding the products of the frequency and far-right column for each row. The covariance is therefore

$$\text{cov}(A, G) = x_{11} - \{x_{10} + x_{11}\}\{x_{01} + x_{11}\}$$
$$= x_{11}(1 - x_{10} - x_{01} - x_{11}) - x_{10}x_{01}$$
$$= x_{11}x_{00} - x_{10}x_{01}.$$

This expression is the same as the one in the main text.

From the above, it is clear that NG always has higher fitness than any other type. Thus if any process breaks up the perfect association between the signal and the altruism gene, nonaltruists who also signal will invade and erode the value of the signal. To see this in another way, consider the covariance between the altruism gene and the green-beard gene:

$$\text{cov}(A, G) = \text{E}(p_i q_i) - \text{E}(p_i)\,\text{E}(q_i)$$
$$= \text{E}(p_i q_i) - pq,$$

where p_i and q_i are the frequencies of the altruism allele and green-beard allele *within* individual i (as in Chapter 3). In Box 5.2, we show that the covariance above is

$$\text{cov}(A, G) = x_{11}x_{00} - x_{10}x_{01}.$$

In population genetics, the covariance between the allele frequency at one locus and the allele frequency at the second is usually labeled D and called *linkage disequilibrium*. When linkage disequilibrium is high, A's are mostly associated with

G's. When A's and G's are assorted at random, $D = 0$. When A's are mostly paired with N's, D is negative.

Recall now one general form of Hamilton's rule that we derived in Chapter 3:

$$b \times \beta(p_j, p_i) > c$$

where

$$\beta(p_j, p_i) = \frac{E(p_j|p_i) - E(p_j)}{p_i - E(p_i)}.$$

We have a similar situation here, but now we are trying to predict the frequency of the altruism allele in the recipient (p_j) given the frequency of the green-beard gene in the recipient (q_j). The same expression will govern selection here, but now

$$\beta(p_j, q_j) = \frac{E(p_j|q_j) - p}{1 - q} = \frac{\dfrac{x_{11}}{x_{11} + x_{01}} - p}{1 - q}.$$

It will be helpful now to re-express the above using only p, q, and D. We show in Box 5.3 how to translate $x_{00}, x_{01}, x_{10}, x_{11}$ into p, q, D. Returning to our expression for β, we obtain

$$\beta(p_j, q_j) = \frac{\dfrac{pq + D}{q} - p}{1 - q} = \frac{D}{q(1 - q)}.$$

Thus the strength of kin selection is proportional to the amount of linkage disequilibrium, D, and inversely proportional to the variance, $q(1 - q)$, of the green-beard allele. Rare markers work best.

Selection can maintain linkage disequilibrium if two alleles are coadapted. In this case, however, selection works against linkage, since NG always has higher fitness than AG. What is left to maintain linkage then? Unfortunately, not much, and it turns out that recombination will eventually unlink all loci if it is unopposed by selection. Everything in this model works against maintaining a strong association between A and G. D will decline until selection does not favor the A allele at all, no matter what it is associated with.

Box 5.3 Representing the four genotypes with p, q, D

Notice that each of our genotype frequencies can be expressed in terms of p, q, and D:

$$x_{10} = p - x_{11}$$
$$x_{01} = q - x_{11}$$
$$x_{00} = 1 - x_{11} - x_{10} - x_{01}.$$

This much is by definition. This implies that

$$D = x_{11}\{1 - x_{11} - (p - x_{11}) - (q - x_{11})\} - (p - x_{11})(q - x_{11}).$$

Simplifying and solving for x_{11}, we obtain

$$x_{11} = pq + D.$$

There is a similar expression for each genotype:

$$x_{10} = p(1 - q) - D$$
$$x_{01} = (1 - p)q - D$$
$$x_{00} = (1 - p)(1 - q) + D.$$

This argument does not apply to genes that generate phenotypic markers that can be used for kin recognition because kin recognition uses a signal to estimate the chance the other individual shares an allele at any locus. Such gene-based systems of kin recognition usually work by identifying a rare marker present in both the actor and recipient. Because it is rare, only closely related individuals are likely to share it, so it may be a reliable cue of kinship.[68]

There is some evidence that single-gene green-beards do exist in some species.[69] In simple chemical cases, on small scales, single genes might be able to produce a marker as well as preferentially direct benefits towards other individuals with the marker. In organisms like ourselves, however, such a mechanism seems unlikely.

5.4   Social learning

The behavior of other individuals is a rich source of information, even when none of it is a consequence of deliberate signaling. As a result social learning is widespread in nature. Social learning is related to signaling because both involve attending to and interpreting information produced by another organism. Furthermore, if you think teaching plays an important role in social learning (and supposedly you do, since you are reading this book), then you also need to worry about how natural selection would design the signals produced by teachers.

There are at least two good reasons people have wondered about the evolutionary origins of social learning. First, people, being consummate imitators, often have the intuition that the animal world is full of social learning. Over the last century, a good deal of evidence has accumulated that social learning probably is common in nature, but that fancy social learning of the human type is quite rare, especially among primates (who are surprisingly bad at social learning). This has led to the obvious questions of how human and animal social learning differ and what kinds of environments favor the evolution of fancy imitation abilities. Second, rather acrimonious debates have raged over the explanatory power of evolutionary thinking when applied to humans. For the most part, social scientists have argued that evolutionary theory is largely irrelevant to their projects because fancy social learning abilities (culture) supersede genetic influences on behavior. Once social learning evolves, the argument goes, new evolutionary dynamics take hold that are not governed by fitness considerations. On the other side of the battlefield, a small group of social scientists with training in evolutionary biology have argued that culture, however it works, must serve genetic interests because it is a product of natural selection. Thus, the argument goes, not only is evolutionary theory not irrelevant to understanding human behavior, but we can even safely ignore culture while

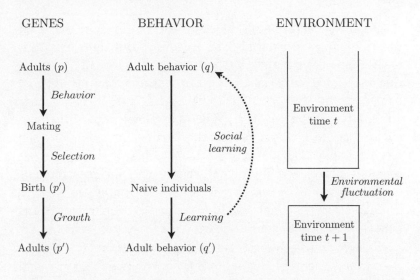

Figure 5.7: The life cycle of the primitive social learning organism, the snerdwump. Different processes change the frequency of alleles for social learning (p), the frequency of adaptive behavior (q), and the state of the environment.

we do so. This has been called the "argument from natural origins."[70]

In order to seriously evaluate arguments like these, we need an evolutionary model in which genes and bodies of socially learned behavior both evolve. Such models are called *dual-inheritance* or *gene-culture coevolution* models because they explicitly treat simultaneous evolution of two separate inheritance systems. In this section, we will introduce a simple dual-inheritance model. Doing so allows us to see both how to model the evolution of learning and how to deal in general with models in which we find multiple simultaneous (or imbedded) recursions. In this case we need one recursion for the genes that lead to the brain architecture that makes social learning possible and another recursion for the distribution of socially learned behaviors in the population. These two recursions will be coupled in some way as well, tying it all together into a sometimes quite messy system.

Begin by assuming that a protocultural species called "snerdwumps" (borrowed from Rogers[71]) live in an environment that occasionally changes states. Specifically, assume that each environmental state has a corresponding optimal behavior, and that when an individual practices that behavior, she receives a payoff b. When practicing some other behavior that is nonoptimal for the current environment, she receives a payoff of zero. Let u be the chance that the environment changes in any given generation. The environment can be in any of a very large number of states, so after it changes there is a new optimal behavior, and all previous behaviors are no longer adaptive.

Snerdwumps acquire their behavior while young. The way that young snerdwumps acquire their behavior is controlled by one of two alleles at a single locus, I or S, which stand for individual or social learning, respectively. Snerdwumps with the I allele acquire their behavior via individual learning. Learning has a cost due to mistakes, accidents, and the time it takes to learn, and we will represent this cost with a reduction in payoff of c units. Suppose that individual learners always acquire the currently adaptive behavior. Then fitness of an individual learner snerdwump is

$$W(I) = w_0 + b - c.$$

Snerdwumps with the S allele save themselves the costs of individual learning by observing what another snerdwump did in the previous generation and imitating that strategy. Social learners cannot guarantee that they acquire the currently optimal behavior, but they pay no learning costs. Let q be the frequency of the adaptive behavior among social learners from the *previous* generation and p be the frequency of social learners. The fitness of a social learner snerdwump is therefore

$$W(S) = w_0 + b(1 - u)\big((1 - p) + pq\big).$$

If the environment has not changed, then socially learned behavior may be adaptive. If the social learner imitates an individual learner (this happens $1-p$ of the time), the behavior acquired is always adaptive. Otherwise (p of the time), she imitates a social learner and acquires adaptive behavior q of the time. Figure 5.7 summarizes the joint genetic, behavioral, and environmental aspects of the snerdwump life cycle.

Natural selection acts on the frequency of social learners, p, and learning processes govern the frequency of the adaptive behavior, q. How do we know the value of q? The recursion for q is simply

$$q' = (1 - u)\{(1 - p) + pq\} + u(0).$$

This follows from the fact that individual learners always acquire the adaptive behavior, and they are imitated $1 - p$ of the time. The behavior learned from one is only adaptive. Next, a social learner is imitated p of the time, and this leads to acquiring the currently adaptive behavior when the environment has not changed and the social learner imitated possessed the adaptive behavior. Immediately after an environmental change none of the behavior is adaptive, and thus $q = 0$. Then, as you can see by looking at Figure 5.8, q converges toward 1 as long as the environment does not change, at which point it is reset to 0. Since the environmental switches occur randomly, the overall time path of q would look something like that shown in Figure 5.9.

Notice that we can now rewrite the fitness of social learners as

$$W(S) = w_0 + bq'.$$

This makes a lot of sense because the value of imitated behavior depends upon what proportion of behavior among adults now will be adaptive when the learner becomes an adult.

This seems like a really hard problem. The recursion for the frequency of social learners, p, is embedded in the recursion for q, which is in turn embedded in the recursion for p. This wouldn't be so bad, but one of them is random. Yikes! What now?

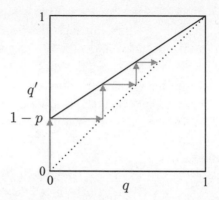

Figure 5.8: A graph of q' against q that shows that q increases at a decreasing rate after an environmental shift.

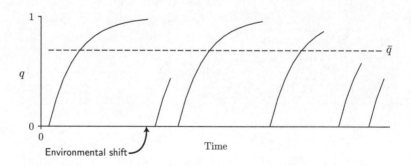

Figure 5.9: A graph of q against time that shows that q increases after each environmental shift, but that environmental shifts occur unpredictably. The long-run expected value of q is $\bar{q}$, shown by the dashed line.

Happily, there is a sneaky way out. The first step is to see that even though the frequency of favorable traits among imitators never settles down to an equilibrium value, the probability distribution does. So after a while the probability of the population taking each value will converge to a constant. This is called the *stationary distribution*. Of course, p will also be changing, and this complicates things. We deal with this problem by assuming that social learning dynamics take place on much shorter time scales than genetic dynamics. This is reasonable in many cases because animals typically undergo many cultural "generations" in each biological generation. Also, the frequency of learned behaviors can change very fast relative to the distribution of genes because whereas learning psychology may be very powerful, the behavior that is learned may have only a small impact on fitness. If social learning dynamics really do act fast enough relative to genetic dynamics, we can assume that gene frequencies are constant and then solve for the genetic change assuming that it responds to the average value of q, $\bar{q}$, over the whole stationary distribution. Each value of p will imply a different equilibrium value of $\bar{q}$, of course, but since social learning dynamics happens on a faster time scale than the dynamics of genes, $\bar{q}$ will instantly reach its new equilibrium anytime p responds to selection. Numerical work shows that this approach gives an excellent approximation if the cultural dynamics are substantially faster than the genetic dynamics.

To find the steady distribution of socially learned behaviors in the population, find the equilibrium value of $\bar{q}$:

$$\bar{q} = (1 - u)\,\mathrm{E}\left\{(1 - p) + pq\right\}$$
$$= (1 - u)(1 - p + p\bar{q}) = \frac{(1 - p)(1 - u)}{1 - p(1 - u)}.$$

We can then compute the fitness of social learners at this

stationary value of q:

$$\hat{W}(S) = w_0 + b\bar{q}$$

$$= w_0 + b\frac{(1-p)(1-u)}{1-p(1-u)}. \tag{5.5}$$

Now we have two closed fitness expressions, and we can proceed as usual. First ask when social learners can invade a population of individual learners:

$$\hat{W}(S) > W(I)$$

$$u < \frac{c}{b}.$$

Provided the environment isn't too unstable, as defined by the relative cost and benefit of individual learning, social learners can invade a population of individual learners. Supposing the above condition is satisfied, could an individual learner invade a population of social learners? We have

$$W(I) > \hat{W}(S)$$

$$b > c.$$

Provided individual learning ever pays, a rare individual learner can invade a population of social learners. Why? Because when the population has no individual learners, social learners cannot track the environment and therefore the accumulated knowledge in the population is useless. Thus a rare individual learner always does better than the rest of the population.

When environments are stable enough that social learners can invade, there are two unstable pure equilibria. This means that there is a stable internal equilibrium at which both individual and social learners exist at some frequency. This equilibrium exists when

$$\hat{W}(S) = W(I)$$

$$\hat{p} = \frac{c - ub}{c(1-u)} = \frac{1 - ub/c}{1-u}. \tag{5.6}$$

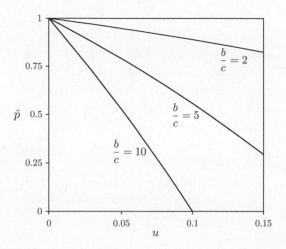

Figure 5.10: The equilibrium frequency of social learning (Equation 5.6). As environments becomes less predictable (u increases), there is less social learning at equilibrium. As individual learning becomes more costly (c increases), there are more social learners at equilibrium.

Figure 5.10 shows how this equilibrium changes as a function of u for three costs of individual learning. As the benefits of practicing the optimal behavior increase, this equilibrium frequency decreases, leading to more individual learners. As the costs of individual learning increase, this point goes the other way, leading to more social learners at equilibrium. As the environment becomes more stable (u approaches zero), this frequency goes to one. The reason is that stable environments allow a greater store of accurate behavior to accumulate among social learners. When the environment shifts, all this stored wisdom becomes useless, so the rarer this occurs, the better social learners do. But when individual learning is relatively costly ($b/c = 2$, for example), substantial amounts of social learning can exist at equilibrium, even when the environment is unreliable.

There are at least two interesting points to make with such a simple model. First, many smart people have argued that the value of social learning is that it saves us the costs of individual experiments and mistakes. The resulting fitness increases should lead species with fancy social learning abilities, like humans, to thrive. But notice that selection in our model here will inevitably lead the population mean fitness to be the same as that of a population of purely individual learners. Such a result doesn't jive well with our intuitions about the value of social learning. Surely, the argument might go, social learning is what has allowed humans to colonize and dominate so many environments while other primates languished in the tropics. The simple model above demonstrates, however, that if social learning increases mean population fitness, it must do so through some method other than saving social learners the costs of individual learning.[72] (This is also a clear demonstration of how natural selection does not necessarily lead to an increase in mean fitness.)

Second, with regard to the debate over whether social learning invalidates the expectation that behavior will be fitness maximizing or can be safely ignored, the model shows that neither position is justified by deduction alone. In principle, the model here can produce both behavior uncoupled from genetic fitness, when the frequency of social learning is high, and behavior tightly linked to our expectations derived from an acultural model, when the frequency of social learners isn't too high. You can see this clearly when considering the equilibrium frequency of adaptive behavior at the equilibrium frequency of social learners ($\hat{p}$). Let A be the expected frequency of adaptive behavior among adults at equilibrium. Then

$$A = \frac{(1 - \hat{p})(1 - u)}{1 - \hat{p}(1 - u)}.$$

Substituting in the value of $\hat{p}$ (Expression 5.6) and simplifying yields

$$A = 1 - \frac{c}{b}.$$

Thus in this model, the frequency of adaptive behavior depends upon the ratio of the costs of individual learning and the benefits of behaving adaptively. The uncertainty of the environment, u, matters as well, but does not appear in this final expression because it affects the equilibrium amount of learning, $1 - \hat{p}$, in such a way that it cancels the reduction in information quality introduced by uncertainty. If the costs of individual learning are high, then at equilibrium a substantial fraction of social learners will be practicing maladaptive behaviors that they acquired from previous generations. This means that the conclusion that we can expect simple social learning to lead to the same behavioral outcomes as individual learning is in principle incorrect. This is additionally perplexing because a population of individual learners would have the same mean fitness, yet nearly everyone in such a population would practice the adaptive behavior (when $\hat{p} \approx 0$, $A \approx 1 - u$). However, if c is small relative to b, then most social learners will behave adaptively, making an acultural model a very good approximation. In principle, then, neither assuming social learning makes no difference nor assuming social learning changes everything is correct *a priori*.[73]

While this model is the simplest we could devise, it turns out to capture the qualitative behavior of a large family of models of the evolution of social learning.[74] It makes no difference, for example, if social learning is also costly, if individual learning is error prone, if environments vary in space as well as time, if social learners can identify and only imitate individual learners, or if individuals use both individual and social learning (rather than distinct strategies). We think this is a good lesson for both sides of such debates to check their philosophical baggage at the door and make serious attempts to understand the dynamic consequences of social learning on behavior. In a species with social learning, you can't understand learned behavior if you focus only on individual psychology. Population dynamics can matter here just as much as they do in the genetic arena.

Guide to the Literature

Signaling theory. The place to begin is Maynard-Smith and Harper's 2003 book *Animal Signals*, which provides a readable and theoretically informed survey of signaling theory. Economists have applied signaling theory to a wide variety of topics since Spence's 1973 Nobel prize–winning work. The first mathematical work in biology came in the late 1980s and was focused on mate choice. The key papers were by Pomiankowski 1987 and Grafen 1990c. We will consider signaling and sexual selection in Chapter 8. **Costly signaling.** Maynard Smith 1991 introduced the discrete Sir Philip Sidney game (also see Godfray 1991), which was soon extended to the continuous case by Johnstone and Grafen 1992. Bergstrom and Lachmann demonstrated the existence of pooling equilibria in the Sir Philip Sidney game and argued that the high cost of separating equilibria makes them unlikely evolutionary outcomes. For a clear and occasionally humorous review of the mathematical theory of signaling among relatives, see Godfray and Johnstone's 2000 survey. **Greenbeards.** Alan Grafen's 1990b paper provides a clear argument on why green beards should be rare. See also Grafen 1998 on the possible existence of green beards. **Social learning.** There is an extensive literature on the evolution of social learning. See Richerson and Boyd 2005 for a nonmathematical introduction and references to the mathematical literature. Very similar models apply to foraging in groups—see Giraldeau and Caraco 2000.

Problems

1. It is always OK to say nothing. In the text, we found the conditions for honest signaling to be an ESS in the simple discrete version of the Sir Philip Sidney game. Assume that the parameters are such that signaling is stable only when it is costly. (a) Show that the strategy pair Always Give, Never Signal is also an ESS in the discrete Sir Philip Sidney game, assuming $d < r\big((1-p)b+pa\big)$. (b) Show that if this condition is not satisfied, then Never Give and Never Signal is an ESS.

2. Sometimes it's better say nothing at all. In the text you learned that the minimum-cost signal that would allow stable, honest communication in the discrete Sir Philip Sidney game had a cost $\hat{c} = b - rd$. (a) Assume that Responder is common in a population and that individuals who are needy signal with cost $\hat{c}$. Compute the average fitness of donors in the population. Compute the average fitness of donors in a population in which the strategy Always Give and Never Signal is the equilibrium behavior. Show that donors in the nonsignaling population have higher fitness if

$$b > \frac{d}{r}(1 - p) + prd.$$

(b) Repeat the exercise in part (a), except now show that beneficiaries have higher fitness than donors if $b > rd$. (c) Show that there are parameter combinations in which both equilibria (signaling and nonsignaling) are stable and in which the average fitness of both donors and beneficiaries is less in the signaling population than in the nonsignaling population. (*Hint*: Plot all three conditions as a function of p and think. This one is tricky.)

3. Cheap talk and conflicts of interest in the battle of the sexes game. People often say that signals must be costly when there are conflicts of interest, at least in one-shot situations. In this problem you will show that this claim is not completely accurate.

In the problems for Chapter 2, you were introduced to the "battle of the sexes" game. Here's a biological version of the game. Suppose that mates forage separately during the day but then rendezvous at night. Usually they prefer to sleep at a dry site because they are safer from predators, but if they have failed to get enough to drink they, prefer a wet site. The probability that they prefer the dry site is $\gamma > 0.5$. If an individual rendezvous with its mate at a preferred site, it gets a payoff of $g > 0$ (for "good"), but if it rendezvous with its mate at the less preferred site, it gets a payoff of b (for "bad," $g > b > 0$). If they fail to rendezvous, they both get zero. (a) First assume that individuals of this species cannot communicate, and show that the strategy always go to the dry site (D) is an ESS against strategies that always go to the wet site (W) and also against a strategy that goes to its preferred site (P). (b) Now suppose that one individual, say the male, can leave a scent marking that points to the site that he will go to. Show that the strategy of "when male, signal your preferred site and then go there, and when female, go to the signaled site" is an ESS against "always go to dry." Assume that females go to their preferred site when there is no signal. Assume a 50:50 sex ratio.

4. External commitment. Tom Schelling, one of the true geniuses of strategic thinking, gave the following example of what is called an "external" commitment device. Suppose that a young woman, call her Sue, has been kidnapped. The kidnapper has contacted her parents, and the ransom has been paid. The kidnapper, Joe, is now faced with the choice of releasing Sue or killing her. Joe is not a sociopath and would prefer not to kill Sue, but because she can identify him, if he does not, he is much more likely to be apprehended and punished. Sue promises not to rat Joe out if she is released. (a) Draw the game tree and use backward induction to show that Joe shouldn't believe her. (b) Schelling makes the following recommendation: Sue, he says, should propose to commit a crime to which Joe is a witness, after which Joe would release her. Draw the new game tree, and use backward induction to show why this act commits Sue to silence, and therefore allows her to survive. Explain what the payoffs must be for Schelling's idea to work.

5. Imperfect individual learning. Modify the social learning model in this chapter to include the fact that individual learning sometimes leads to acquiring a nonoptimal behavior. Specifically, assume that each individual learner has a chance e of acquiring a nonoptimal behavior. Show that the equilibrium frequency of social learners is now

$$\hat{p} = \frac{1 - (1 - e)ub/c}{1 - u}.$$

Explain how the rate of errors in individual learning affects the equilibrium frequency of social learners.

Notes

[53]There are several ways to cut up communication. We endorse no particular taxonomy, but see Hurd and Enquist 2005 for one recent classification.

[54]Tinbergen 1952.

[55]Dawkins and Krebs 1978.

[56]Maynard Smith 1976.

[57]Hirshleifer 1977.

[58]Maynard Smith 1991.

[59]Bergstrom and Lachmann 1997, 1998.

[60]Johnstone and Grafen 1992 analyzed a similar model. Here we follow the treatment of Bergstrom and Lachmann 1997.

[61]Bergstrom and Lachmann 1997, 1998; Lachmann and Bergstrom 1998.

[62]Bergstrom and Lachmann 1998.

[63]Brilot and Johnstone 2002.

[64]Johnstone 2002.

[65]Silk et al. 2000.

[66]Johnstone 2002.

[67]Dawkins 1976 invented the clever label. Hamilton 1964 seems to have come up with the idea, however.

[68]Grafen 1990b.

[69]See Grafen 1998.

[70]Boyd and Richerson 1985.

[71]Rogers 1988.

[72]Boyd and Richerson 1995.

[73]Rogers 1988.

[74]Boyd and Richerson 1995. See also Giraldeau and Caraco 2000 for very similar models applied to foraging behavior.

Box 5.4 Symbols used in Chapter 5

Sir Philip Sidney model

p Probability beneficiary severely wounded

d Cost to donor of giving resource

a, b Costs to severely and mildly wounded beneficiary, respectively, of not receiving resource

c Cost to beneficiary of signaling

Continuous Sir Philip Sidney model

y, x Fitnesses of donor and beneficiary, respectively, before social interaction

$c(x)$ Cost to beneficiary of signaling condition x

a Signaling threshold

f Fraction of donors who transfer resource when signaled to

Repeat interaction model

I, i Payoffs to actor and recipient, respectively, when they interact

$E, -e$ Payoffs to actor and recipient, respectively, when they fight

$D, -d$ Payoffs to actor and recipient respectively, when recipient flees

p Chance actor is peaceful

w Chance pair continues to another interaction

Altruism and signaling

x_{ij} Frequency of genotype ij

p Frequency of altruism allele, $x_{11} + x_{10}$

q Frequency of marker allele, $x_{11} + x_{01}$

D Covariance between altruism and marker alleles

Box 5.5 Symbols used in Chapter 5, continued

Social learning

b Payoff of behaving optimally

c Cost of learning individually

u Chance that the environment changes in a given generation

p Frequency of social learning allele

q Frequency of adaptive behavior

A Frequency of adaptive behavior at equilibrium

Chapter 6

Selection among Groups

Ask a nonbiologist to explain some feature of animal behavior and, more likely than not, he will tell you how the behavior ensures the survival of the population, or the species, or even the ecosystem. Without parental care, the species would go extinct. Without pollination, neither the fig wasp nor the fig tree would survive. Nowadays evolutionary biologists have little use for such good-for-the-species explanations. Until the late 1960s, however, such explanations were also commonplace in biology. The sea change came with the publication of *Animal Dispersion in Relation to Social Behavior*, in which the author, ornithologist V. C. Wynne-Edwards, argued that many enigmatic bird behaviors functioned to prevent over-population. For example, some species engage in spectacular group displays near their roosts at sunset. Wynne-Edwards thought that this behavior allowed birds to census their population and thereby decide how many eggs to lay. Wynne-Edwards knew that selection on individuals would favor animals who behaved selfishly.[75] But he thought that populations which lacked reproductive restraint were more likely to go extinct, and that this would overcome the countervailing force of selection on individuals.[76] The book

223

generated a storm of controversy, and luminaries like George Williams and John Maynard Smith penned critiques explaining why this mechanism, then called "group selection," could not work. The result was the beginning of an ongoing and highly successful revolution in our understanding of the evolution of animal behavior, a revolution that is rooted in carefully thinking about the individual and nepotistic functions of behavior.

More recently, however, the widespread acceptance of the Price covariance equation has led to renewed controversy. In Chapter 3, we used the Price equation to derive Hamilton's rule. In this chapter, we will see that the Price equation also leads to a description of natural selection as going on in a series of nested levels: among genes within an individual, among individuals within groups, and among groups. While this *multilevel selection* approach and the older gene-centered approaches are mathematically equivalent, the multilevel approach has proven to be very useful for understanding many evolutionary problems. However, it has also led to confusion about what kinds of evolutionary processes should be called "group selection." Some people use "group selection" to mean the process that Wynne-Edwards envisioned—selection between large groups made up of genealogically distantly related individuals. Others use the term to refer to selection involving any kind of group in a multilevel selection analysis, including pairs of individuals interacting in, say, the Hawk-Dove game.

In this chapter, we will take a general look at *multilevel selection*. We'll see that the effect of selection can *always* be decomposed into effects at different levels and can always be conceptualized as acting only at the individual level—the so-called *personal fitness* approach. Moreover, these approaches are completely equivalent. So if you do your sums properly, both lead to the same answer. Altruism will not evolve in a multilevel model if it does not evolve in an equivalent individual selection model. Different approaches may be more or less useful, depending on the problem (and the modeler).

The real scientific question about group selection is what kinds of population structures and selective regimes allow the evolution of traits that benefit groups. We hope to convince the reader who is unaccustomed to this approach that this way of thinking clarifies the pervasive confusion about "group selection" as well as provides valuable insights into the evolution of social behavior that are not so easy to see when selection is viewed as an individual-level or gene-level process.

6.1 Three views of selection

Population genetics models are a lot like accounting. We create a set of books that keep track of gains or losses of copies of different alleles from one generation to the next, much as accountants add up expenditures and remittances to find net profits (or losses) in a firm. There are three common ways that evolutionary game theorists and population biologists organize their accounts. To keep things simple, let's consider a very simple example. Consider a population with four individuals subdivided into two pairs (Figure 6.1). Individuals in each pair interact in an additive prisoner's dilemma, with benefit $b = 3$ additional offspring and cost $c = 1$ fewer offspring. During this generation, $r = 1$ within each pair. The first pair is made up of two cooperators, and the second is made up of two defectors.

Personal fitness. The simplest way to account for changes in the frequencies of alleles is to sum up the number of offspring of individuals. Each type of individual has an expected chance of survival and produces an expected number of offspring. The product of survival and reproduction is the expected individual fitness of a given type (as in Chapter 1). In the population in Figure 6.1, the average fitness of altruists (black circles) is $w_0 + \{(b-c) + (b-c)\}/2 = w_0 + 2$. The average fitness of nonaltruists (open circles) is $w_0 + \{0 + 0\}/2 = w_0$.

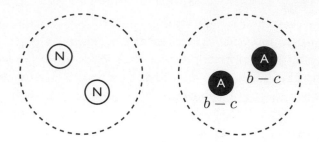

Figure 6.1: A population of four organisms (small circles) in two interacting pairs (large dashed circles). Pairs in this population always comprise clones ($r = 1$). Labels on each individual represent net changes in fitness after social behavior. Altruists (black circles) provide aid worth $b = 3$ additional offspring to the other individual in their pair, at a cost $c = 1$ offspring to themselves. Nonaltruists (open circles) may receive but never provide aid.

The average fitness advantage of altruists is $w_0 + 2 - w_0 = 2$. Note that we are able to ignore relatedness here because we are just counting surviving offspring.

Inclusive fitness.   Another way to think about fitness effects is to credit all effects of social behavior to the gene that caused them. Does an allele, by causing a behavior, produce additional copies of itself through the reproduction of bodies other than its own? As described in Chapter 3, this method leads to Hamilton's rule. It is powerful, but subtle and easily misapplied. If you help your neighbor, your *inclusive fitness* is your change in fitness as a result of the behavior in question, plus your neighbor's change in fitness as a result of your aid, times the coefficient of relatedness between you and your neighbor. Be careful to count only *changes* in fitness that result from your behavior. Do not count changes in your fitness that result from the actions of your neighbor. Let's apply this logic to the population in Figure 6.1. The average inclusive fitness of an altruist is $w_0 - c + br = w_0 - 1 + 3(1) = w_0 + 2$. The inclusive fitness of nonaltruists is similarly $w_0 + 0 + 0r = w_0$. Again, the fitness advantage of altruists is $w_0 + 2 - w_0 = 2$. Here, we need to include relatedness because it expresses the statistical association between types. This example may make the approach look easy (or not), but correctly applying inclusive fitness accounting is tricky.[77]

Multilevel selection. A third way is to decompose selection into its effects in a system of nested levels. Which levels depends on the problem. For example, if individuals interact in pairs, the decomposition could be: among individuals within pairs and among pairs. However, there may also be selection among genes within an individual, due, for example, to meiotic drive, or among larger groups within a species. Selection can have countervailing effects at different levels. At the individual, within-group level, helping in a prisoner's dilemma is detrimental: an individual who aids another in such a situation is never *personally* better off than another in the same group who withholds aid. Every additional altruism allele in an individual reduces its relative fitness, holding the genotypes of its neighbors constant. However, at the group level, groups of altruists have higher average fitness than groups of nonaltruists. Every additional altruist increases group fitness.

We saw in Chapter 3 that selection is driven by the covariance between fitness and allele frequency. In the hierarchical perspective, we compute the covariance at each level of population structure. In Figure 6.1, there are no fitness differences within groups. The covariance between individual fitness and individual allele frequency *within* groups must be zero because there is no variation within groups (they comprise clones, recall). If there is no variance, there can be no covariance. This means that there is no selection within groups.

Selection between groups is summarized by the covariance between average group fitness and average group allele frequency. In this case, the average fitness of the left group is $w_0 + 0$. The average fitness of the right group is $w_0 + 2$.

The average frequency of altruists in the left group is 0. The frequency in the right group is 1. The covariance is

$$\frac{((0)(0) + (2)(1))}{2} - \frac{(0 + 2)}{2}\frac{(0 + 1)}{2} = \frac{1}{2}$$

Selection between groups is positive and favors altruism.

What justifies this kind of decomposition? Why is the covariance at each level relevant, and how do we compare them? In the rest of this chapter, we show you how to use the Price equation to answer these questions.

6.2   Deriving the Price equation

In Chapter 3, we met George Price's covariance genetics and the Price equation. However, our "derivation" wasn't very rigorous, and we promised we'd get serious later. In this section, we derive the Price equation properly, and, in the process, you will see that in its general form, the Price equation has an additional, very important term, a term that is crucial for understanding multilevel selection.

Consider a population of particles of many different types. These could be genes or something else that makes copies of itself. They might also be molecules in a mixture or kinds of cars in California. The particles may come in many variants, but we will keep track of just one of them, labeled S. These particles can be grouped into subpopulations labeled $g = 1, 2, 3, \ldots$. Within each subpopulation the particles themselves are labeled $i = 1, 2, 3, \ldots i$. Thus, the ith particle in subpopulation g is labeled ig, and w_{ig} copies of it are made in a given unit of time, a generation. Now define the terms in the following table:

	Subpopulation g	Entire population
Number of particles	n_g	N
Frequency of S	p_g	p
Mean number of copies	$w_g = \frac{1}{n_g}\sum_i w_{ig}$	$\bar{w} = \sum_i \frac{n_g}{N} w_g$

We add a prime to each variable to denote the value of that variable in the next generation.

With these definitions, it is shown in Box 6.1 that the change in frequency of S is given by

$$\bar{w}\Delta p = \sum_g \frac{n_g}{N} w_g(p_g - p) + \sum_g \frac{n_g}{N} w_g(p'_g - p_g). \qquad (6.1)$$

Recall that the definition of covariance is $\text{cov}(x, y) = \text{E}(xy) - \text{E}(x)\text{E}(y)$. After a bit of algebra, this can be rewritten as

$$\text{cov}(x, y) = \text{E}\left\{x\left(y - \text{E}(y)\right)\right\}.$$

Thus the first term on the right side of Equation 6.1 is a covariance and the second is an expectation:

$$\bar{w}\Delta p = \text{E}\left(w_g(p_g - p)\right) + \text{E}\left(w_g(p'_g - p_g)\right)$$
$$= \text{cov}(w_g, p_g) + \text{E}(w_g \Delta p_g). \qquad (6.2)$$

Expression 6.2 is the general form of the Price equation.

When we used the Price equation to derive Hamilton's rule in Chapter 3 our "subpopulations" were individuals. Individual organisms are collections of genes. Thus the number of "particles" in a diploid individual "subpopulation" is always 2. On this interpretation, the expectation on the right side of Expression 6.2 is the average change in allele frequency *within* individuals in the population. In most organisms, most of the time, this is approximately equal to zero because Mendel's laws ensure that each gene has a fair chance of making its way into gametes. There are forces like mutation and meiotic drive that sometimes cause $\Delta p_g > 0$. For the most part, such effects can be ignored, however, and that is why we could do all our work in Chapter 3 without the second term on the right side of Expression 6.2.

Now we need to include this term. To see why, let's write down the Price equation thinking of subpopulations as groups of individuals, not groups of genes within individuals. Let g be the index of the gth group in the population. Looked at

---

Box 6.1 Deriving the Price equation, the nitty gritty

Using the definitions in the text, we find that the frequency of S
after one generation is

$$p' = \sum_g \frac{n'_g}{N'} p'_g.$$

The frequency in the whole population is the sum of the number
of individuals with S in each group, $n'_g p'_g$, over all groups divided
by the population size N'. Thus the new frequency in each subpopulation is weighted by the number of individuals in it, and this
gives us the weighted average frequency of S in the population as
a whole. Now notice that $n'_g = w_g n_g$ and $N' = \bar{w}N$. The new
number of copies is just the number of particles times the mean
number of copies produced per particle. And so

$$p' = \sum_g \frac{w_g n_g}{\bar{w}N} p'_g.$$

Let's subtract p from both sides and multiply by the mean fitness
$\bar{w}$ to make a difference equation:

$$\bar{w}\Delta p = \sum_g \frac{n_g}{N} w_g p'_g - p\bar{w} = \sum_g \frac{n_g}{N} w_g p'_g - p\left\{ \sum_g \frac{n_g}{N} w_g \right\}.$$

Now we simultaneously add and subtract $\sum_g \frac{n_g}{N} w_g p_g$ and then
factor. We'll label each of the four terms, so it is easier to see
where they go. **C** and **D** are the terms we add. We have

$$\bar{w}\Delta p = \overbrace{\sum_g \frac{n_g}{N} w_g p'_g}^{\textbf{A}} - \overbrace{p\sum_g \frac{n_g}{N} w_g}^{\textbf{B}} + \overbrace{\sum_g \frac{n_g}{N} w_g p_g}^{\textbf{C}} - \overbrace{\sum_g \frac{n_g}{N} w_g p_g}^{\textbf{D}}$$

$$= \left(\overbrace{\sum_g \frac{n_g}{N} w_g p_g}^{\textbf{C}} - \overbrace{p\sum_g \frac{n_g}{N} w_g}^{\textbf{B}} \right)$$

$$+ \left(\overbrace{\sum_g \frac{n_g}{N} w_g p'_g}^{\textbf{A}} - \overbrace{\sum_g \frac{n_g}{N} w_g p_g}^{\textbf{D}} \right)$$

$$= \sum_g \frac{n_g}{N} w_g (p_g - p) + \sum_g \frac{n_g}{N} w_g (p'_g - p_g).$$

in this way, the Price equation says that the change in the frequency of the allele is equal to the covariance between the frequency of the allele in group g and the mean fitness in group g plus the expectation (average) of the product of the change in allele frequency in group g and the mean fitness in group g.

Now here's the nifty part. The product $w_g \Delta p_g$ in the expectation looks like the product $\bar{w} \Delta p$ on the far left. On the far left, we have the average fitness in the population times the change in the average frequency of the allele in the population. On the right, we have the average fitness in group g times the change in frequency of the allele in group g. Thus the Price equation tells us that

$$w_g \Delta p_g = \text{cov}(w_{ig}, p_{ig}) + \text{E}(w_{ig} \Delta p_{ig}),$$

where i is the index of individuals within groups. This is the same as our previous interpretation of subpopulations with new notation. This equation says that the average fitness times the change in allele frequency in group g is equal to the covariance between the allele frequency in individual i in group g and the fitness of individual i in group g, plus the expectation of the fitness of individual i in group g times the change in allele frequency in individual i in group g.

Again, let's assume that there is no meiotic drive and that mutation is weak enough to ignore. This means that $\Delta p_{ig} = 0$. Then we can combine our expressions for $\bar{w} \Delta p$ and $w_g \Delta p_g$ to yield

$$\bar{w} \Delta p = \text{cov}(w_g, p_g) + \text{E}\left(\text{cov}(w_{ig}, p_{ig}) + \text{E}(w_{ig} \Delta p_{ig})\right)$$
$$= \text{cov}(w_g, p_g) + \text{E}\left(\text{cov}(w_{ig}, p_{ig})\right).$$

This is the full expression for the change in the frequency of the allele for a population structured into groups. It is very important to understand that we haven't changed anything but how to do the fitness accounting. Before, we ignored population structure and averaged across individuals. Now we first average across individuals within groups and then across groups within the population. If we do the math correctly, both approaches will yield the same answer. Note also that we could, in theory, perform the decomposition of the Price equation any number of times to have highly complex structured models of groups within groups within groups.

6.3   Selection within and between groups

The form of Price's equation that we have just derived provides some important insight right away. Recall that we can always express a covariance as a variance times a regression coefficient:

$$\text{cov}(x, y) = \text{var}(x)\beta(y, x).$$

Thus we can express the Price equation as

$$\bar{w}\Delta p = \text{var}(p_g)\beta(w_g, p_g) + \text{E}\left(\text{var}(p_{ig})\beta(w_{ig}, p_{ig})\right).$$

The beta coefficients tell us the effect of selection at each level of population structure. That is, $\beta(w_{ig}, p_{ig})$ is the regression of individual fitness on individual allele frequency. Similarly, $\beta(w_g, p_g)$ is the regression of group fitness on group allele frequency. Altruism is a puzzle because these two numbers have opposite sign. Altruism increases group fitness but decreases individual fitness, relative to defection. Thus one factor which determines whether altruism evolves is the magnitude of these two regression coefficients.

The other important variables are the variances at each level of population structure. Many readers will already be familiar with analysis of variance (ANOVA), a common statistical model which partitions variance into within and between group components. The form of the Price equation performs a similar partitioning. If most of the variance in the population is within groups but all groups have nearly the same frequency of altruism alleles, then $\text{var}(p_g)$ will be very small, whereas the expectation of $\text{var}(p_{ig})$ will be nearly the entire variance in the population. Thus, in order for altruism to evolve, there must be variation among groups in the frequency of altruism alleles.

6.3.1 Inter- and intrademic group selection

We are now almost ready to explain why Wynne-Edwards was wrong. However, first we need to define the endpoints of a continuum of group selection processes.[78] At one end of the continuum is *interdemic* group selection. Here, groups are demes, meaning that they are large, genetically well-mixed populations that are partially isolated from each other. Genetic drift, the founder effect, and sometimes selection can give rise to variation among such groups because these groups are partially isolated. At the other end of the continuum is *intrademic* group selection. Here, the groups are not isolated genetically, and members of different groups frequently mate. There may be strong assortative interaction, and therefore substantial variation among groups, but it must be due to some kind of behavioral assortment. There are clearly many situations which are intermediate between these extremes.

Wynne-Edwards thought that large, very long lived, partially isolated groups with more reproductive restraint were less likely to go extinct than groups in which the genes causing reproductive restraint were less common. Therefore, genes for behaviors leading to reproductive restraint would spread. Thus, he thought that interdemic group selection could lead to the evolution of altruism.

Wynne-Edwards was mostly wrong because $\text{var}(p_g)$ is usually quite small. To see why, consider the evolutionary forces which affect the relative amount of variation between groups.

Mutation. Mutation increases the variation in the population groups. Mutations are like a trickle of water, filling the bucket of variation. Selection and drift both remove variation and sort it among and within groups.

Selection. Altruists are selected against in every group, and this means that natural selection acts to reduce the variation among groups.

Migration. The movement of individuals from one group to another tends to make all groups similar. This reduces the variation between groups.

Genetic drift. This force increases the variation among groups if different groups drift to different frequencies of the altruism allele. The sampling variation that arises when new groups arise is a from of genetic drift. Genetic drift may be especially important when new groups are founded by small numbers of individuals.

Thus the only force generating variance among groups is genetic drift. We know that drift is very weak unless populations are quite small. It also has to compete against migration, which is a usually strong force in large mobile animals. Consider, for example, that in many mammals and birds, all members of one sex disperse from their natal group. Such mixing will quickly eliminate genetic variation between groups. A classic theoretical result is that two subpopulations are unified with respect to drift if[79]

$$2Nm > 1,$$

where N is the number of individuals in the total population and m is the migration rate as a proportion of individuals per generation. Since we can express m as the number of migrants M over the total population size N ($m = M/N$), this condition becomes

$$M > \frac{1}{2}.$$

If there is more than one migrant every other generation, significant variance cannot evolve between the subpopulations! This result is what economists call a "stylized fact," a conclusion that glosses over many complications, is held by no one

Table 6.1: Estimates of the ratio of the variation between groups to the variation within groups from two small scale human populations. The Yanomama are a South American horticultural/foraging population that live in small, seminomadic villages at very low population densities. The Gainj and the Kalam are two small-scale horticultural groups that live in the fringe highlands of northeastern Papua New Guinea. Because these estimates are based on variation at a small number of marker loci, they are likely to be upper bounds on the relative strength of selection among groups on the altruistic genes because such genes will necessarily be under selection. (From D. Hartl and A. Clark, *Principles of Population Genetics*, 3rd ed., Sinauer, Sunderland, MA, pp. 300–301).

Population	Number. of loci	$\frac{\mathrm{var}(p_g)}{\mathrm{var}(p_{ig})}$
Yanomama	7	0.0434
Gainj/Kalam	5	0.0230

to be an actual description of the world, but nevertheless tells us something valuable. The above result depends on the assumption that drift and migration have reached equilibrium, an assumption that may often be violated. Nonetheless, it captures the essential insight that it is very hard to maintain variance between groups when individuals migrate between groups.

This is why Wynne-Edwards–style group selection is not usually a source of altruism in nature. Returning to the Price equation, if $\mathrm{var}(p_g) \approx 0$,

$$\bar{w}\Delta p \approx \mathrm{E}\{\mathrm{var}(p_{ig})\beta(w_{ig}, p_{ig})\}.$$

Thus, the change in the frequency of the altruism allele will depend almost entirely upon the effect of selection within groups, no matter how much altruism increases group survival. To show that this isn't all just theory, Table 6.1 shows some (gasp) data on the amount of variation within and among groups for two small scale human populations. These data say that the effect of a gene on the group welfare will have to be between 25 and 50 times greater than its effect on individual welfare for the two selection forces to balance.

Of course, if migration is weak enough, variation among groups can persist long enough for selection among groups to act on it. In some species of invertebrates, this appears to be the case.[80] In large primate groups, however, it does not.

6.3.2   Kinds of interdemic group selection

To get beyond this intuitive (aka arm-waving) argument, we've got to specify some more details. Interdemic models are more complex than the intrademic models so far in two ways:

1. *Migration pattern:*   Interdemic models must specify how migration is limited. That is, interdemic models must specify the pattern of migration among groups. Two simple models are commonly used:

   - Island model migration: Each generation, a fraction m of each group (or "subpopulation") emigrates and is replaced by individuals drawn at random from all other groups.

   - Stepping stone model migration: Here populations are arranged in space in a ring (a one-dimensional stepping stone) or on a torus[81] (a two-dimensional stepping stone). Subpopulations only exchange migrants with their nearest neighbors.

   Migration patterns in real populations presumably lie somewhere between these two extremes.

2. *How groups compete.*   In intrademic models, groups that produce more individuals leave more genes in the subsequent generation. There are two common ways groups can compete in interdemic models:

   - Differential proliferation models: Groups with a higher frequency of altruists produce more emigrants. In such models intergroup competition is limited by the rate of mixing between groups. Low migration rates and stepping stone migration facilitate the generation of variation among groups but also limit the possibility of group competition.

- Differential extinction models: Groups with higher frequencies of altruists are less likely to go extinct than groups with few altruists. Extinct habitats are recolonized by members of surviving groups. If all the colonists are drawn at random from the population as a whole, called the migrant pool model, recolonization also destroys variation among groups. If, on the other hand, colonists are drawn from a single, randomly chosen subpopulation, called the propagule pool model, colonization leaves the variation among groups unchanged. Differential extinction models suffer from the fact that the rate of extinction of large groups is probably low for most species.

There have been numerous models that mix and match these features. For example, Alan Rogers[82] analyzed a one-locus diploid differential proliferation model. Rogers assumed that variation among groups comes to equilibrium quickly compared to changes in gene frequency. The average juvenile in group i produces $\alpha(1 + gp_i)$ emigrants, where p_i is the frequency of the (weakly) altruistic allele A. Thus selection among groups favors A. How much depends on how many of these emigrants find their way to other subpopulations, which in turn depends on the frequency of altruists in the population as a whole. As altruists become more common, there will be more emigrants, and thus more competition for a fixed number of slots. Rogers showed that the mean fitness of group i is

$$w_i = 1 + b(p)(p_i - p),$$

where p is the frequency of A in the population and

$$b(p) = \frac{mg}{1 + gp}.$$

The survivorship of adults is given by

$$\begin{array}{ccc} aa & Aa & AA \\ 1 & 1-c & 1-2c \end{array}$$

Thus, selection within groups favors a, the nonaltruist allele. Rogers then showed that the altruistic gene will increase in the population if

$$\frac{c}{b(p)} < \frac{\text{var}(p_i)}{\text{E}\{\text{var}(p_{ij})\}}.$$

So far, this is just a restatement of Price's formula for the particular form of intergroup competition that Rogers assumed. To go further, he assumed that both individual and group selection are weak enough that their effects on the variation among groups can be ignored. With this assumption, he could use the neutral model to calculated the variances within and among groups. According to this approximiation, the altruistic gene will increase whenever

$$\frac{g}{c} > \frac{2n}{\Omega}.$$

In intrademic models in which groups are formed at random, $\Omega = 1$. In contrast, if groups were made up of full-sibs, $\Omega = 2n$. This provides a natural scale on which to judge the effectiveness of interdemic selection. If Ω is near one, interdemic group selection is no more effective than intrademic group selection with random group formation, which is to say, it cannot lead to the evolution of strong altruism. If Ω is large, then interdemic group selection is effective.

Rogers showed that for island model migration at equilibrium

$$\Omega = \frac{K-1}{K(2-m)},$$

where K is the number of groups and m is the migration rate. Thus, $\Omega < 1$ for reasonable parameter values! With island model migration, interdemic group selection is *less* effective than intrademic group selection with random group formation. The reason is that in the intrademic models groups are reformed each generation, and this pumps new variation into the metapopulation. Here, migration and other forces slowly leach away variation among groups. Rogers also calculated Ω for various stepping stone models, and unless migration rates are very low, $\Omega < 2$.

Kenichi Aoki[83] analyzed a differential extinction model in the same spirit. Aoki assumed that the rate of group extinction is proportional to $1 - kp_i$, where p_i is the frequency of A in group i, but that A is selected against within groups just as in Rogers' model. Assuming propagule pool recolonization and weak selection, he derived the following condition for A to increase:

$$\frac{k}{c} > 2nm.$$

Assuming $k = g$ *sensu* Rogers' model, this is less stringent than in the differential proliferation model. Also, the strength of group selection increases as m gets smaller, while in Rogers' model the strength of group selection decreases as migration increases. In the differential proliferation model migration has two competing effects. First, it reduces variation among groups, but second, it allows more group competition. In the differential extinction model only the first effect is present, so group selection can be made strong simply by decreasing m. Notice, however, that m still has to be on the order of $\frac{1}{n}$ for the strength of group selection to equal the strength of individual selection.

6.3.3 Kin selection as intrademic group selection

In this section, we consider a multilevel approach to evolution in a population structured in pairs with coefficient of relatedness r. If our claim that the individual-level and multilevel approaches are equivalent, then we should recover Hamilton's rule, $rb > c$. Let's see if that's true.[84] As we have already seen, the Price equation tells us that selection will favor altruism when

$$\text{var}(p_g)\beta(w_g, p_g) + \text{E}\{\text{var}(p_{ig})\beta(w_{ig}, p_{ig})\} > 0.$$

The variance terms tell us how much variation there is among and within groups and summarizes the structure of the population. The regression coefficients describe the direction and magnitude of selection at each level. In the remainder of this section we derive expressions for each of these terms.

Selection between groups. First, let's compute the regression of group fitness on group genotype, $\beta(w_g, p_g)$. The easiest way to do this is to draw a picture (Figure 6.2). With every additional altruist, the average fitness in the group rises $(b - c)/2$ units. Since $b > c$, more altruists mean higher average fitness. The regression coefficient is the slope of the line passing through the points in Figure 6.2. This slope is the rise, $b - c$, over a unit change in average group genotype. Thus $\beta(w_g, p_g) = b - c$. At the group level, selection favors altruism.

Selection within groups. To compute the regression of individual fitness on individual genotype, $\beta(w_{ig}, p_{ig})$, a similar picture helps (Figure 6.3). To do this correctly, we have to consider how a change in an individual i's genotype, p_i, affects individual fitness, w_i, holding the genotype of the other individual constant. Consider a group with two altruists. If one becomes a nonaltruist, then his fitness is $w_0 + b$, while the altruist's fitness is $w_0 - c$. In a group with two nonaltruists, if one switches to altruism, we arrive at the same fitness difference. Thus the line passing through the points in Figure 6.3 defines the regression coefficient $\beta(w_{ig}, p_{ig})$. Since this line sinks $b + c$ units over a unit change in genotype, the slope is $-(b + c)$. Selection within groups works against

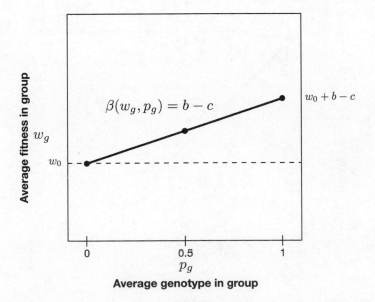

Figure 6.2: Graph of average individual fitness within a group as a function of average individual genotype within a group ($p_g = 0$ is no altruists, $p_g = 0.5$ is one altruist, $p_g = 1$ is two altruists). Although an altruist is always worse off than a nonaltruist in the same group, groups with more altruists have higher average fitness. The slope of the line through the fitness points is the regression coefficient $\beta(w_g, p_g)$. As average fitness rises $b - c$ units with one unit of genotype, the slope is $b - c$. Since $b > c$, this is positive, and selection favors altruism at the group level.

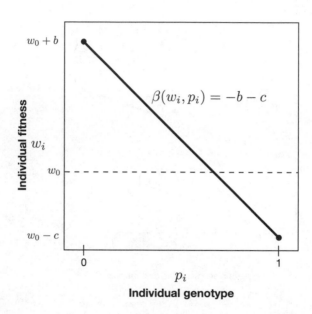

Figure 6.3: Graph of individual fitness as a function of individual genotype ($p_i = 0$ is a nonaltruist, $p_i = 1$ is an altruist). In a pair with one altruist and one nonaltruist, the altruist has fitness $w_0 - c$ and the nonaltruist has $w_0 + b$. The slope of the line through these two points is the regression coefficient $\beta(w_i, p_i)$. As fitness falls $b + c$ units as genotype goes from 0 to 1, the slope is $-(b + c)$. Selection acts against altruism at the individual level.

altruism because an individual is always worse off by switching her behavior to altruism.

Variance between groups. Increasing the coefficient of relatedness within the pair increases the variance between groups because individuals within groups will be more similar to each other. However, the exact expression depends upon the total variance in the population. To compute the variance between groups, $\text{var}(p_g)$, we need to recall the expressions presented in Chapter 3 that define how kin-biased association affects probabilities of interaction. Each of those probabilities specifies the chance that different kinds of pairs form. Across many groups, these probabilities tell us the expected proportion of the population consisting of each kind of group. Using these values together with the p_g values each kind of group creates, Box 6.2 tells us that the variance between groups is

$$\text{var}(p_g) = \text{E}(p_g^2) - \text{E}(p_g)^2 \tag{6.3}$$

$$= \frac{1}{2}p(1-p)(1+r). \tag{6.4}$$

As r increases, more of the variance is between groups, until, when $r = 1$, the variance between groups is $p(1-p)$, the total variance in a haploid population. When $r = 0$, assortment is random, and half of the variance is between groups.

Variance within groups. Like the variance between groups, the variance within groups is a function of the amount of assortment within pairs, as well as the global frequency of altruism alleles. It differs however, because we are concerned with the expected (average) variance within groups. We already computed the regression of individual gene frequency on within group fitness, and this is a constant across pairs,

---

Box 6.2 Computing the variance between groups

The probabilities of each pair (group) forming arise from the haploid kin-selection model presented in Chapter 3:

$$\Pr(A, A) = p(r + (1 - r)p)$$
$$\Pr(A, N) = 2p(1 - p)(1 - r)$$
$$\Pr(N, N) = (1 - p)(r + (1 - r)(1 - p))$$

where $\Pr(A, N)$ is the chance a group with one altruist (A) and one nonaltruist (N) forms, p is the frequency of the altruism allele in the population, and r is the average coefficient of relatedness within pairs. Each of these pairs implies a value of p_g. Finding the variance is cognitively less taxing if we use a table to lay out the computation:

Pair	Probability of pair	p_g	p_g^2
A, A	$p(r + (1 - r)p)$	1	1
A, N	$2p(1 - p)(1 - r)$	0.5	0.25
N, N	$(1 - p)(r + (1 - r)(1 - p))$	0	0

The expectation of p_g is p, by definition. The expectation of the square is

$$E(p_g^2) = \Pr(A, A)(1) + \Pr(A, N)(0.25) + \Pr(N, N)(0)$$
$$= p(r + (1 - r)p) + \frac{1}{2}p(1 - p)(1 - r).$$

The variance (defined by Expression 6.3) is therefore

$$\mathrm{var}(p_g) = p(r + (1 - r)p) + \frac{1}{2}p(1 - p)(1 - r) - p^2.$$

After factoring and simplifying, this becomes Expression 6.4 in the main text.

---

so we can factor it out to yield

$$E\{\mathrm{var}(p_{ig})\beta(w_{ig}, p_{ig})\} = -(b + c)\, E\{\mathrm{var}(p_{ig})\}.$$

This allows us to compute the expected variance instead of the expected product of the variance and regression coefficient. In Box 6.3, we show that the expected variance within

---

Box 6.3 Computing the average variance within groups
The expected variance within groups is just the weighted average of the variances of each kind of pair:

$$E\{var(p_{ig})\} = Pr(A, A) \, var(p_{ig}|p_g = 1)$$
$$+ Pr(A, N) \, var(p_{ig}|p_g = 0.5)$$
$$+ Pr(N, N) \, var(p_{ig}|p_g = 0).$$

Since the variance in genotype within pairs with two altruists or two nonaltruists is zero (no variance in types means no variance)

$$E\{var(p_{ig})\} = Pr(A, N) \, var(p_{ig}|p_g = 0.5)$$
$$= 2p(1 - p)(1 - r) \, var(p_{ig}|p_g = 0.5).$$

Use a table to help compute the variance:

Individual	Frequency	p_{ig}	p_{ig}^2
A	0.5	1	1
N	0.5	0	0

Following the definition of variance, we have

$$var(p_{ig}|p_g = 0.5) = E(p_{ig}^2) - E(p_{ig})^2$$
$$= (0.5)(1) + (0.5)(0) - (0.5)^2 = 0.25.$$

Using the fact that $Pr(A, N) = 2p(1 - p)(1 - r)$, we obtain the expected variance within groups as in to be Expression 6.5 in the main text.

---

groups is

$$E\{var(p_{ig})\} = \frac{1}{2}p(1 - p)(1 - r). \tag{6.5}$$

Summing this expression and the expression for the variance between groups, Expression 6.4, shows that they do in fact add up to $p(1 - p)$, the total variance in the population. As r increases, less variance exists within groups because individuals within pairs are more similar to one another. When $r = 1$, Expression 6.5 goes to zero because individuals within pairs are always the same type.

Reassembling the Price equation.   Let's put the whole thing together now:

$$\bar{w}\Delta p = \text{var}(p_g)\beta(w_g, p_g) + \text{E}\{\text{var}(p_{ig})\beta(w_{ig}, p_{ig})\}$$
$$= \frac{1}{2}p(1-p)(1+r)(b-c) - \frac{1}{2}p(1-p)(1-r)(b+c)$$
$$= p(1-p)(rb-c).$$

Thus the frequency of the altruism allele increases when $rb > c$, which is Hamilton's rule. The derivation works because the multilevel selection framework—the strategy of partitioning selection into within- and between-group components—is mathematically equivalent to the older, individual-centered accounting. There is only one world out there. It would be bad if changing the way we did the accounting of genes changed the answer.

The fact that we can express any case of multilevel selection as a case of selection on individual alleles within bodies does not mean that the gene's-eye view is the "right" way to think about selection. Likewise, the fact that we can derive Hamilton's rule in a multilevel selection framework does not mean that inclusive fitness is the "wrong" way to think about selection. The choice of which way to build the model should be based upon how useful each approach is for a given problem.[85] In many cases, the inclusive fitness or personal fitness approach is best. It is possible, for example, to reframe reciprocal altruism (Chapter 4) in terms of multilevel selection. Consider that TFT always does worse than ALLD within a pair. It is only in pairs of TFT that prolonged cooperation yields payoffs that can compensate for the negative within-group effects. However, we think it is easier to model reciprocal altruism the way we did in Chapter 4, using personal fitness.

These derivations assumed dyads. It is useful to know how much variance will exist between groups of arbitrary size n. The derivation of these variances is tedious, so we'll just skip to the result (but see Box 6.4, if you like tedious). In general, if all groups consist of n individuals of average relatedness r

Box 6.4 Deriving variance terms for groups of size n

Begin with the definition of the variance among groups, then insert the definition of p_g:

$$\text{var}(p_g) = \text{E}\left\{(p_g - p)^2\right\}$$

$$= \text{E}\left\{\left(\frac{1}{n}\sum_i p_{ig} - p\right)^2\right\} = \frac{1}{n^2}\text{E}\left\{\left(\sum_i (p_{ig} - p)\right)^2\right\}.$$

To understand what we are about to do, consider that, if we have the three values x_1, x_2, and x_3 and we square the sum of these, we get an expansion, $(x_1 + x_2 + x_3)^2 = x_1x_1 + x_1x_2 + x_1x_3 + x_2x_1 + x_2x_2 + x_2x_3 + x_3x_1 + x_3x_2 + x_3x_3$. As a square diagram, we see the diagonal terms that are squares and off-diagonal terms that are products of different values:

	x_1	x_2	x_3
x_1	x_1^2	x_1x_2	x_1x_3
x_2	x_1x_2	x_2^2	x_2x_3
x_3	x_1x_3	x_2x_3	x_3^2

The x's are like the values $(p_{ig} - p)$ in our problem. We are squaring a sum of these, so the geometry is the same as the x example just above. This lets us pull out the diagonal terms:

$$\text{var}(p_g) = \frac{1}{n^2}\text{E}\left\{\sum_i (p_{ig} - p)^2 + \sum_i \sum_{j \neq i} (p_{ig} - p)(p_{jg} - p)\right\}.$$

The first sum is the total squared deviations from the population mean. If we take the expectation of this sum, it will almost be the population variance, $\text{var}(p_i)$. However, it is missing a factor $1/n$, so instead it is $n \times \text{var}(p_i)$. The expectation of the second summation is almost the covariance between genotypes of pairs of interacting individuals. Again, however, we are missing factors. This time we are missing factors of $1/n$ and $1/(n-1)$. All together, we now have

$$\text{var}(p_g) = \frac{1}{n^2}\left\{n\,\text{var}(p_i) + n(n-1)\,\text{cov}(p_i, p_j)\right\}$$

$$= \text{var}(p_i)\left\{\frac{1}{n} + \frac{(n-1)}{n}\frac{\text{cov}(p_i, p_j)}{\text{var}(p_i)}\right\}.$$

The term $\frac{\text{cov}(p_i, p_j)}{\text{var}(p_i)}$ happens to be our definition of relatedness, r (see Chapter 3). Because variances add, $\text{E}(\text{var}(p_{ig}))$ can be found by subtracting the expression for $\text{var}(p_g)$ from $\text{var}(p_i)$.

and var(p_i) is the total variance in the population:

$$\text{var}(p_g) = \text{var}(p_i) \left\{ \frac{1}{n} + \frac{n-1}{n} r \right\}$$

$$\text{E} \left\{ \text{var}(p_{ig}) \right\} = \text{var}(p_i) \left\{ \frac{n-1}{n} (1-r) \right\}.$$

If groups are formed at random ($r = 0$), then a proportion $1/n$ of the total variance is between groups. For pairs, this isn't so bad, but for groups of 10, this is already very small.

6.3.4 Selection between multiple stable equilibria

There are good reasons to be skeptical of interdemic group selection models of the evolution of altruism. However, there is one sort of interdemic group selection that is not nearly as controversial—selection among multiple stable equilibria. In Chapter 4, we saw that models of the evolution of reciprocity all had many different stable equilibria, some at which reciprocators are common and others at which defectors are common. In Chapter 5 we saw that models of signaling also had many different stable equilibria. And this is only the tip of the iceberg. Many games generate multiple stable equilibria. In fact, the famous folk theorem of economic game theory states that any sufficiently iterated game will have a large number of such equilibria.

When there are multiple stable equilibria, interdemic group selection can be a powerful force even when groups are large and migration rates substantial. To see why, let's apply the Price equation to this problem. To make things concrete, think about the iterated n-person prisoner's dilemma we analyzed in Chapter 4. There were two strategies, ALLD and T_{n-1}, the strategy that cooperates first and continues to cooperate as long as all others cooperated. Suppose that the population is structured with island model migration, and in some subpopulations ALLD is common, while in others T_{n-1} is common. Remember that both strategies are ESSs, meaning that they are favored by selection when common. Thus, if selection is strong compared to migration between groups, each strategy will remain common in some groups. Within each group the common type will be favored, but only weakly because the subpopulation is near equilibrium. However, groups at the cooperative equilibrium will have much higher average fitness. Thus

$$\bar{w}\Delta p = \underbrace{\text{var}(p_g)}_{\approx\text{all variance}} \underbrace{\beta(w_g, p_g)}_{>0} + \text{E}(\underbrace{\text{var}(p_{ig})}_{\approx 0} \underbrace{\beta(w_{ig}, p_{ig})}_{\approx 0}).$$

Interdemic group selection doesn't usually lead to the evolution of altruism because drift doesn't generate enough variation among groups unless groups are small or migration rates are very low. However, when there are multiple equilibria, selection *increases* variation among groups, and if selection is strong, there will be lots of variation at equilibrium. As usual, selection among groups favors strategies that increase average fitness, but because selection within groups is weak, the between group predominates. This means that group selection will determine which strategy (or strategies) spread throughout the population in the long run. This process is sometimes referred to as equilibrium selection, and it has an interesting literature of its own.[86]

6.4   Dispersal

The evolution of dispersal provides an example of how multilevel selection can be used to clarify an important biological problem. Most organisms disperse—at some point in their lifecycle they leave their birthplace and move somewhere new. The world outside is unknown and dangerous. Finding a new niche is a risky proposition. Why not just stay home? Inbreeding avoidance is one reason. In many species dispersal reduces the chance that close kin mate. A second reason is that dispersal allows organisms to escape poor habitats and colonize more favorable ones. W. D. Hamilton and Robert

May[87] showed that dispersal can be favored even if inbreeding is not an issue and the environment is entirely uniform and unchanging. When there is density-dependent competition, selection among groups favors dispersal despite the cost to individuals.

Imagine an asexual organism that lives in an environment with n sites which organisms can occupy. At each site only one adult can survive. Thus, there is local density dependent population regulation in each site because no matter how many juveniles/seedlings are produced, only one can occupy the site as an adult. Assume that all individuals produce k offspring and then die. Of these k offspring, a proportion v disperse. Dispersal is risky, and as a consequence only a fraction p of dispersers survive. Surviving dispersers are then each equally likely to settle at any site in the environment. $(1 - v)k$ offspring remain at each site, and these offspring compete with immigrants and with one another.

To calculate the ESS (CSS) value of v, suppose that most of the population creates vk dispersers, but one mutant produces $k(v + \delta)$ dispersers and $k(1 - v - \delta)$ home-bodies. We assume δ is a small number. We want to determine the value of v that prevents the mutant type from invading. When the population is entirely one type and they are all competing on equal ground for n sites, each adult will on average have one successful offspring because local population regulation ensures only one adult can exist at each site. Thus, to find the ESS value of v, we determine the value at which the expected number of sites occupied by the mutant type is less than one. (This is another good example of how natural selection does not have to lead to increased fitness at equilibrium.)

Let's begin by considering just a single territory at a time. Figure 6.4 illustrates the flow of individuals in and out of a territory. Think of the n territories as existing along a line. At any habitable point, a resident individual produces some home-body offspring who do not disperse, as well as some footloose dispersers. The home-bodies stay behind to compete with incoming dispersers from other sites, while

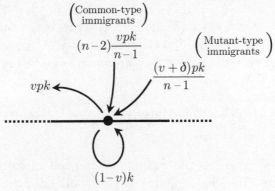

(a) Common-type resident

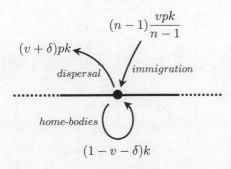

(b) Mutant-type resident

Figure 6.4: The flow of individuals in and out of territories in the Hamilton-May model of dispersal. At both sites occupied by common types (a) and the one site occupied by a mutant (b), the inflows of dispersers from all other sites and outflows of dispersers from the home site produce a pool of competitors for the single adult slot at the site in the next generation. One of the competitors will, at random, become the resident adult.

a fraction of the resident's dispersers survive and compete in other territories. Thus in any given territory, immigrants from other sites and natal home-bodies all compete to be the new resident.

These inflows and outflows (Figure 6.4) imply that the chance that a mutant (M) succeeds at a site previously occupied by an individual of the common type (C) is

$$\Pr(M|C) = \frac{\dfrac{(v + \delta)pk}{n - 1}}{(1 - v)k + \dfrac{(v + \delta)pk}{n - 1} + \dfrac{(n - 2)vpk}{n - 1}}. \qquad (6.6)$$

The numerator is the number of mutant offspring reaching the site. The denominator is the total number of offspring competing at the site, which includes home-bodies, mutant dispersers, and common-type dispersers from the other $n - 2$ sites. Now calculate the expected number of successful mutant offspring across *all* common sites by multiplying 6.6 by $n - 1$, the number of sites with common-type residents. After some simplification, we obtain

$$E(\#M|C) = \frac{(v + \delta)p}{(1 - v) + vp + \dfrac{\delta p}{n - 1}}.$$

Now we need to compute the chance that a mutant replaces its parent at the home site. This is (from Figure 6.4)

$$\Pr(M|M) = \frac{(1 - v - \delta)k}{(1 - v - \delta)k + \dfrac{(n - 1)vpk}{n - 1}}.$$

There is only one of these territories, so $E(\#M|M) = \Pr(M|M)$.

Now we're going to assume that there are a huge number of territories in the environment ($n \to \infty$). Doing this allows us to simplify both $E(\#M|C)$ and $E(\#M|M)$:

$$E(\#M|C) \approx \frac{vp + \delta p}{1 - v + vp}$$

$$E(\#M|M) \approx \frac{1 - v - \delta}{1 - v - \delta + vp}.$$

Then we add these expressions to get the total number of successful mutants at all sites:

$$E(\#M) = E(\#M|C) + E(\#M|M)$$
$$= \frac{vp + \delta p}{1 - v + vp} + \frac{1 - v - \delta}{1 - v - \delta + vp}. \qquad (6.7)$$

Remember that the evolutionarily stable value of v must satisfy

$$E(\#M) < 1.$$

This is so because any mutant that has on average less than one successful offspring will not be able to replace the common type. In Box 6.5, we show that the CSS dispersal proportion is

$$v^\star = \frac{1}{2 - p}. \qquad (6.8)$$

Recall that p is the chance that a disperser survives to reach a site. Consider that when survivorship is small, $p \approx 0$. Then selection favors sending almost half of your offspring out, even though they face almost certain death. The reason is that only one offspring can succeed at your own site. Thus the within-group component of selection gives very poor odds to nondispersers, while the between-group component of selection gives better odds, since potentially all dispersers can survive and succeed. You might object that sites cannot be "groups" because there is only one adult at each. But this misses the point. The k offspring produced at each home site constitute a group, and this pool of offspring is the group relevant to hierarchical selection, just as pairs of kin are the groups in Hamilton's rule.

There is a meaningful analogy here to the altruism models we have already studied. Dispersal is a kind of individually

---

Box 6.5 Deriving the CSS dispersal proportion

We proceed just as we did in Chapter 2 and find the CSS value of v, v^*, algebraically. Substituting the expression for $E(\#M)$ (Expression 6.7) into the condition for stability $E(\#M) < 1$ and cross-multiplying gives

$$(vp + \delta p)(1 - v - \delta + vp) + (1 - v - \delta)(1 - v + vp)$$
$$< (1 - v + vp)(1 - v - \delta + vp).$$

Now assume that δ is small, so that terms containing a factor $\delta^2 \approx 0$. Using this assumption, we see that this condition simplifies, after much joyless algebra (expand and simplify, as usual), to yield

$$\frac{\delta}{2 - p} < \delta v.$$

As in Chapter 2, imagine first that $\delta > 0$. When this is true, this condition is satisfied when

$$v > \frac{1}{2 - p}.$$

This means that when the mutant sends out more dispersers ($\delta > 0$), it will be less successful than the common type when $v > 1/(2-p)$. If v is less than this value, mutants that create more dispersers are favored. When $\delta < 0$, a mutant that sends out fewer dispersers, the common type is stable provided that $v < 1/(2 - p)$. When v is greater than this value, mutants that send out fewer dispersers are favored. Thus the only un-invadable value of v, v^*, is Expression 6.8 in the main text.

---

costly behavior because it reduces an individual's chance of surviving to adulthood. However, by dispersing, an individual avoids competing with its $k - 1$ relatives, enhancing their average fitness. Thus the between-group component of selection favors dispersal, while the within-group component favors being a home-body.

In this model, all siblings are clones, so the variance within groups of kin is zero, leading the between-group component of fitness to dominate the ESS outcome. Hamilton and May also solved a model in which they considered dispersal in a sexual species. They assumed that all males migrate, mate, and die (what a life). The lone individual left at a site is a female. In this model, selection favors less dispersal because offspring are less related to one another in a sexual species. This means that the variance within groups is greater, enhancing the effect of within-group selection, which favors remaining at home. In fact, they found that if disperser mortality is high enough, selection will favor zero dispersal.

Guide to the Literature

Multilevel selection. Price's 1970 paper introduced the covariance equation. His 1972 paper applied it to altruism, as did W. D. Hamilton in a rarely read 1975 essay. D. S. Wilson has made frequent use of this approach in modeling social behavior. His book with Eliot Sober, *Unto Others* (1998), provides a brief for the multilevel approach. **Interdemic group selection**. Wade 1978 reviewed the early literature and introduced the distinction between interdemic and intrademic group selection. Bergstrom 2002 is a useful review from the perspective of an economist. Coyne et al. 1997 reviewed and critiqued Wright's shifting balance theory. This paper provoked spirited defenses of Wright by Wade and Goodnight 1998 and Peck et al. 1998. Aviles' 2002 work on social spiders showed that interdemic group selection is sometimes important in nature. **Dispersal.** Hamilton and May's model stimulated a great deal of work on dispersal in both ecology and animal behavior. Most behavioral ecologists, however, think first of Greenwood's 1980 paper on sex-biased dispersal. See Perrin and Goudet 2001 for a recent model of that argument (also see other chapters in the same volume for reviews of other aspects of dispersal).

Problems

1. When the strength of selection among groups varies

Recall the basic stag hunt game from the homework in Chapters 3 and 4. A population is divided into a large number of pairs of individuals, formed at random. There are two strategies, Hunt Stag (S) and Hunt Hare (H). Hunting hare is a solitary behavior, and always yields a small payoff h. Hunting stag is a cooperative behavior, and if both individuals in the pair do so, each gets a large payoff $s > h$. If either hunts hare, however, the lone stag hunter always gets zero (0). **(a)** Let p be the frequency of S in the population. Using the tools we introduced in Chapter 2, find all the equilibria in this game and determine their stability. **(b)** Think now of the population as being broken up into randomly formed groups of two. Let p_g be the frequency of S in group g and w_g be the average fitness of group g, and let p_{ig} be the frequency of S in individual i in group g and w_{ig} be the fitness of individual i in group g. Compute $\operatorname{cov}(w_g, p_g)$ and the expectation of $\operatorname{cov}(w_{ig}, p_{ig})$. **(c)** Using Price's formula, interpret how selection acts on the different components of this system. When does the between-group component of selection favor S? When does it favor H?

2. Covariance reciprocity

We saw in this chapter how to derive Hamilton's rule using the full Price equation and in Chapter 4 how to model reciprocal altruism from an individual-fitness perspective. Using the covariance method, rederive the basic stability criterion $wb > c$ for TFT to be an ESS against ALLD. Think of pairs as groups formed of individuals related by an average coefficient of relatedness r. Use the Price equation to compute the condition for p, the frequency of TFT in the population, to increase. What component of fitness favors TFT?

3. Local dispersal and the Hamilton-May model

In most species, dispersing individuals are more likely to settle near their natal area than in more distant areas. Dispersal of seeds by the wind or by animals provides important examples. In this problem you will determine how the effect of distance on dispersal modifies the conclusions of the Hamilton-May model. Assume that the sites in the Hamilton-May model are arranged in a ring and that the fraction of individuals who disperse all land in one of the two sites that adjoin the site of their parent. Otherwise, the model is as described in the text, individuals are asexual, a fraction p of dispersers survive, each adult produces k offspring, and the number of sites is large. Derive the ESS dispersal fraction. Explain why your result makes sense.

4. Sex specific selection on genes on the X chromosome

Some years ago a geneticist at the National Institutes of Health thought he had discovered a gene on the X chromosome that predisposed males to be homosexual. Subsequent research has cast doubt on his findings, and it is now thought to be unlikely that this gene exists. If it did, and accounted for any substantial fraction of males who adopt a homosexual lifestyle, it would present a problem for evolutionary biology because such a gene would be strongly selected against. One possible solution to the puzzle is that the same gene has a positive effect on the fitness of women. (Bob Trivers suggested that human males are so obnoxious that such a gene might be necessary for women to be sexually attracted to males.) Obviously, it would be of interest to know how much of an advantage for females is necessary to balance a given amount of selection against the gene in males. The fact that the gene is on the X chromosome, and therefore males carry one copy and females two copies, makes this tricky. Solve this problem using the Price covariance formula. Suppose that:

1. Half of the population is male and half is female

2. Males with one copy of the G allele have fitness $1 - h$ and males with no copies have fitness 1.

3. Females with two copies of the G allele have fitness 1, GN females have fitness $1 - \frac{1}{2}d$, and NN females have fitness $1 - d$

4. Mating is at random, so that female genotypes can be calculated using Hardy-Weinberg proportions.

Calculate the change in frequency of the G allele.

Notes

[75] Page 18 of the 1962 edition of Wynne-Edwards *Animal Dispersion in Relation to Social Behavior* makes this clear.

[76] Page 20 of his 1962 book: "Some [communities] prove to be better adapted socially and individually than others, and tend to outlive them, and sooner or later to spread and multiply by colonising the ground vacated by less successful neighboring communities."

[77] See Grafen 1984 and Queller 1996.

[78] These distinctions were first introduced by Michael Wade 1978.

[79] Wright 1931.

[80] Aviles 1993, 2002.

[81] A torus is a doughnut-shaped surface.

[82] Rogers 1990.

[83] Aoki 1982.

[84]This type of derivation is due to Price 1972 and Hamilton 1975. Wade 1985 demonstrated how the Price equation can be used to address many seemingly different types of selection.

[85]Several people have made this point. See Dugatkin and Reeve 1994, for example.

[86]Sewall Wright discussed such interdemic group selection as part of his "shifting balance" theory. See Coyne et al. 1997 for an evaluation and references. Boyd and Richerson 1990, 2002, analyzed cultural evolution models based on differential extinction and differential proliferation, respectively.

[87]Hamilton and May 1977.

Box 6.6 Symbols used in Chapter 6

b Benefit of being the recipient of altruism

c Cost of behaving altruistically

r Coefficient of relatedness between actor and recipient (see Chapter 3)

n Number of sites adults might occupy (in dispersal model)

v Proportion of offspring who disperse

k Average number of offspring

p Chance each offspring survives dispersal

δ Change in dispersal proportion of mutant adult's offspring

Chapter 7

Sex Allocation

When your second author (R. B.) was a graduate student at the University of California Davis, one of his professors, W. J. Hamilton III,[88] used to always pop the same question on qualifying exams:

> *Why in most species are there equal numbers of males and females?*

This was before the theoretical revolution in behavioral ecology, and most students had no idea how to frame a proper answer. Most started down the road of an even sex ratio being somehow good for the species. Then Bill would pounce. "Every farmer knows that only a few males are needed to do the job," he would say, "so why 50 percent males?" More than one graduate student spiralled into a panic, never to return. Nowadays students are much more sophisticated, but in our experience, this question still creates problems. This is a shame because the answer to Bill's question is one of evolutionary biology's greatest theoretical triumphs.

In this chapter we explain why a 50:50 sex ratio is so common, as well as why sex ratios sometimes deviate from 50:50. We will see that sex ratio is really about appreciating the ways that parents invest in sons and daughters. There are deep reasons why, under most circumstances, mothers should invest the same amount in their sons and daughters. There

261

are also interesting exceptions to this rule. And as you may have come to expect, this chapter provides an opportunity to introduce two new tools, the calculus of reproductive value and (at the end of the chapter) matrix methods that extend equilibrium analysis to dynamic systems with two or more state variables. Both of these approaches are valuable and, once mastered, will allow you to model evolutionary dynamics in more complex systems than we have seen so far. Once you understands these tools, you will be ready to go beyond this text and tackle more challenging problems such as selection in age-structured populations.

7.1   Fisher's theory of sex allocation

The answer that Hamilton was looking for was first sketched by R. A. Fisher in his 1930 book[89] and goes like this.

- Every individual has a mother and a father. (Hard to argue with that.)

- This means that the total reproductive success[90] of males as a whole and females as a whole must be the same.

- If there are fewer males, then *each* male will have higher reproductive success than each female. The reverse is true if females are less common.

- This means in turn that any female who makes fewer of one sex will do worse than one who has equal numbers of sons and daughters.

Like many of Fisher's arguments, this one is both subtle and very terse, and as a result its hard to be sure that it is correct. Here's the problem: As long as sex ratio doesn't affect family size, the reproductive success of females who have equal numbers of sons and daughters and those who make an uneven number is exactly the same. It is only among grand-offspring that the 50:50 females do better. To see this, suppose that the average number of grand-offspring produced per daughter is R and that most mothers in the population produce f daughters and m sons. Such a female's fitness is

$$W(f,m) = fR + m\frac{f}{m}R = 2fR.$$

Each son she produces has a chance f/m of mating each daughter in the population. Since the common type produces f daughters and m sons, this is also the population sex ratio any given son or daughter faces when finding a mate. Similarly, a mutant mother who produces $f^\star = f + d$ daughters and $m^\star = m - d$ sons has fitness

$$W(f^\star, m^\star) = (f + d)R + (m - d)\frac{f}{m}R.$$

Notice that the mutant faces the same population sex ratio as the common type. Using the method introduced in Chapter 2 for solving CSS problems (see Box 7.1), we see that the common type is stable against all invaders, provided $f = m$. That is, the number of daughters must equal the number of sons, or else some other sex ratio could invade.

This simple model seems to confirm Fisher's intuition. But what is special about the second generation? Why stop there? Couldn't the advantage increase in great-grand-offspring or vanish in great-great-grand-offspring? Why was counting offspring ("reproductive success") adequate up to this point, but suddenly it won't suffice? We need a more principled approach that explains why counting offspring does not work in this case, why it worked up to this point, and why the 50:50 sex ratio is the right answer.

7.2 Reproductive value and Fisherian sex ratios

The key to solving this problem involves another idea of Fisher's, the concept of *reproductive value*. In all of the models so far in this book, we have implicitly assumed that the

---

Box 7.1 Fisherian sex ratio as a CSS problem

The common type is stable provided $W(f, m) > W(f^\star, m^\star)$. This implies

$$2fR > (f + d)R + (m - d)\frac{f}{m}R.$$

Dividing both sides by R, multiplying by m, and reducing, we obtain

$$0 > d(m - f).$$

If $d > 0$, this is true if $f > m$. That is, if the mutant produces more daughters, the common type is an ESS if it produces more daughters than sons. No type that produces even *more* daughters can invade. Conversely, if $d < 0$, the common type is an ESS only if $f < m$. Thus the only globally stable family composition is $f = m$.

---

only difference between individuals is their inherited strategy. In real life, individuals differ in many ways—they may be male or female, young or old, dominant or subordinate—and these differences can have a profound effect on their ability to reproduce. Moreover, inherited strategies can vary in the types of individuals they produce as well as the number. When this is the case, it won't do to just count offspring. We need some way of weighting the reproductive value of different types.

In his 1930 book, Fisher defined *reproductive value* as the expected relative contribution of an individual of a given type to the population, now and in the future.[91] It is *expected* because it is a statistical aggregate of different likely and unlikely outcomes. It is *relative* because we scale reproductive value against the reproductive value of individuals of different ages or types. It includes both present and all future reproduction. Reproductive value is thus a number that tells us the relative contribution of an individual to all future generations. We can use this number to rescale benefits and costs among different classes of individuals.

So let's use this idea to analyze the evolution of sex ratio. Let v_f be the per capita reproductive value of a female and v_m be the reproductive value of a male. Let n_m be the number of males in the population and n_f be the number of females. In a diploid species like ourselves, fathers and mothers contribute equally to the genetic composition of offspring (sex chromosomes and organelles aside). This means that the sum contributions of males and females to the future gene pool of the population must be equal. Thus the total reproductive value of all the males must equal the total reproductive value of all the females, or

$$v_f n_f = v_m n_m.$$

If instead females contributed 2/3 of the genes to each offspring and males contributed 1/3, the reproduction of females would be twice as valuable as the reproduction of males, and $v_f n_f = 2 v_m n_m$. Such species do exist,[92] but we'll stick with a diploid example. Now solve for v_f:

$$v_f = \frac{n_m}{n_f} v_m. \tag{7.1}$$

Thus, the value of daughters relative to sons depends on the population sex ratio. It is this strong frequency dependence that drives the evolution of sex allocation. When males are more numerous than females, $n_m/n_f > 1$, each daughter contributes relatively more to future generations (has higher reproductive value) than does each son. When females are more numerous, males are more valuable. Since reproductive values are relative, this is all we need to know about them in this case.

Using the fact that the number of offspring an individual produces should be weighted by the reproductive value of each, we show in Box 7.2 that a mutant female who produces $f^\star$ daughters and $m^\star$ sons will invade a population of females producing f daughters and m sons, provided

$$\frac{f^\star}{f} + \frac{m^\star}{m} > 2. \tag{7.2}$$

Box 7.2 Deriving the Shaw-Mohler theorem

Weighting offspring by their reproductive value, we find that the fitness of a common-type mother who produces f daughters and m sons is

$$W(f,m) = \frac{1}{2}\left(f \cdot v_f + m \cdot v_m\right).$$

The factor $1/2$ accounts for the fact that only half of the mother's genes are present in each offspring. The fitness of a mutant mother who produces $f^\star$ daughters and $m^\star$ sons is

$$W(f^\star, m^\star) = \frac{1}{2}\left(f^\star \cdot v_f + m^\star \cdot v_m\right).$$

Substitute the reproductive value of females, Equation 7.1, into the fitness expressions of both types of mother:

$$W(f,m) = \frac{1}{2}\left(f \frac{n_m}{n_f} v_m + m \cdot v_m\right)$$
$$W(f^\star, m^\star) = \frac{1}{2}\left(f^\star \frac{n_m}{n_f} v_m + m^\star \cdot v_m\right).$$

Now, notice that the population sex ratio, n_m/n_f, is equal to the expected sex ratio within common-type families, m/f. Thus

$$W(f,m) = m \cdot v_m$$
$$W(f^\star, m^\star) = \frac{1}{2} v_m \left(f^\star \frac{m}{f} + m^\star\right).$$

The mutant invades provided that $W(f^\star, m^\star) > W(f, m)$. We have

$$\frac{1}{2} v_m \left(f^\star \frac{m}{f} + m^\star\right) > m \cdot v_m$$
$$f^\star \frac{m}{f} + m^\star > 2m.$$

Dividing both sides by m yields Expression 7.2 in the main text.

Box 7.3 Equal sex ratio via the Shaw-Mohler theorem

Define $f^\star = f + d$ and $m^\star = m - d$. Then substituting these values into the Shaw-Mohler theorem (Expression 7.3) gives

$$\frac{f + d - f}{f} + \frac{m - d - m}{m} > 0$$

$$\frac{d}{f} - \frac{d}{m} > 0$$

Multiplying both sides by $m \cdot f$ and simplifying yields the condition

$$d(m - f) > 0$$

for the mutant producing $f^\star$ daughters and $m^\star$ sons to invade. An argument like that in Box 7.1 shows that the only ESS family composition in this case is one in which $f = m$.

Equivalently,

$$\frac{f^\star - f}{f} + \frac{m^\star - m}{m} > 0. \tag{7.3}$$

Expression 7.3 is known as the *Shaw-Mohler theorem*, after the first people who derived it.[93] The derivation here assumed the simplest possible life history—two sexes, no age classes, no overlapping generations. However, Expression 7.3 applies approximately to a large number of more-general life histories as well.[94]

7.3 Using the Shaw-Mohler theorem

What does this theorem tell us? One obvious lesson is that any mutant that increases both the number of sons and the number of daughters will be favored by selection. You hardly needed all this math to tell you that. The real use of the Shaw-Mohler theorem comes when there are limits that force mothers to trade off the number of sons and daughters, so that one of the two terms is positive and the other negative. Then the Shaw-Mohler theorem tells us exactly how selection

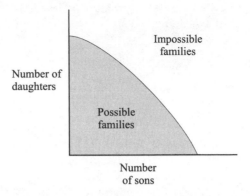

Figure 7.1: The fitness set gives the feasible combinations of sons and daughters. Each point in the plot is a potential family composition. However, females are constrained by a "budget" that allows them to produce only families within the shaded region, the *fitness set*. All potential families to the right and above this region are outside a female's budget.

makes the tradeoff. First, if we assume again that males and females are equally costly so that adding one son necessitates subtracting one daughter, we recover the simple Fisherian result. At equilibrium, $f = m$ (Box 7.3). If we relax this constraint, a couple of more-general implications emerge.

To see what they are, we need to introduce a way of formalizing the tradeoff. One way to do this is with a fitness set as shown in Figure 7.1. The two axes are the number of sons and daughters in a family. The shaded area plots the families that are possible, given the resource constraints that bear on mothers. The Shaw-Mohler theorem says that the ESS family has to lie on the outside boundary, called the "frontier," of the fitness set because for any point inside the fitness set there are mutants that increase both the number of sons and that of daughters. The Shaw-Mohler theorem also tells us exactly where along the boundary the ESS lies. To see how, let's rewrite the theorem in terms of $f^\star$, f, $m^\star$, and m:

$$\frac{f^\star - f}{f} + \frac{m^\star - m}{m} > 0$$

$$f^\star - f > -\left(\frac{f}{m}\right)(m^\star - m).$$

The second expression is the equation for a line that goes through the point (f, m) with slope $-(\frac{f}{m})$. So to find out if a combination (f, m) is an ESS, we plot the line through that point and see if there are any feasible mutants that can invade. For example, suppose we choose a point on the frontier of the fitness where the common-type females produce more sons than daughters. This leads to the graph shown in Figure 7.2(a). All the points in the darkly shaded area have values of $f^\star$ above the line and can invade. Notice that the slope of the line is low because $(\frac{f}{m})$ is less than one. So let's try a potential ESS with more females than males. This results in the plot shown in Figure 7.2(b). Now the slope of the ESS line is greater than one, and mutants that lead to more males can invade.

By now you might be able to see that the ESS family composition is going to be the point on the boundary where the slope of the line tangent to the fitness set is equal to $-(\frac{f}{m})$, as is shown in Figure 7.2(c). Because the line is tangent to the fitness set, there are no mutants that are possible and satisfy the tradeoffs imposed by the Shaw-Mohler theorem. If there is an algebraic expression for the fitness set frontier of the form $f = F(m)$ and if this function is continuous without any kinks, then you can find the ESS sex ratio by taking the derivative of F with respect to m, setting it equal to $-(\frac{f}{m})$, and solving. For example, suppose the fitness set is defined by $f = F(m) = b - m$, where b is a constant ("brood" size). The derivative of this function with respect to m is $dF/dm = -1$. Setting this equal to $-(\frac{f}{m})$ yields the ESS sex ratio. You will see what happens if the function has kinks when you solve Problem 7.4.

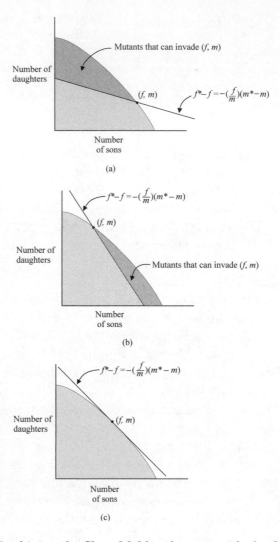

Figure 7.2: Combining the Shaw-Mohler theorem with the fitness set. (a) Points in the darkly shaded portion of the fitness set are possible and yield higher fitness in a population with the sex ratio (f, m). Because there are more males than females, mutants that lead to more daughters can invade. (b) When there are more females than males, mutants that lead to more males can invade. (c) When the fitness set is convex, the point at which the line defined by the Shaw-Mohler theorem is tangent to the boundary of the fitness set gives the ESS sex ratio.

Now back to Fisher. Suppose that making male babies and female babies consumes the same amount of maternal time, effort, energy, and other resources so that the frontier of the fitness set is an straight line with slope -1. Then the ESS sex ratio is equal number of males and females. You should try out some other points and see why they are not ESSs. Now it is often the case that the offspring of one sex are cheaper to produce, meaning that females can make more of them than they can of the other sex. In mammals females are usually cheaper than males because males are larger and require more food, create longer interbirth intervals, etc. In many invertebrates, females are more expensive. As is shown in Figures 7.3(a) and 7.3(b), you can use the Shaw-Mohler theorem to show that the ESS sex ratio is biased in favor of the cheaper sex so the total investment in each sex is equal.

All of the fitness sets you have seen so far are convex, meaning that a straight line between any two points on the boundary lies inside the set. Some readers may be wondering what happens when the fitness set is concave. This would occur when it pays to specialize—making all of one sex is cheaper than making a mix. Then it seems like the only ESSs would be all males or all females, and clearly that won't do. The answer is that you can extend the fitness set, as shown in Figure 7.4, by making families that are all males and all females in the proportions given by the relative costs of such families. For example, half of females make all sons and half make all daughters.

7.4 Biased sex ratios

We saw just now how the Shaw-Mohler theorem will sometimes predict producing more of one sex at equilibrium. Such equilibria are still inherently Fisherian because total investment in each sex is equal. However, there are a variety of situations that can lead to more investment in sons than daughters or vice versa. We consider two cases: (1) when

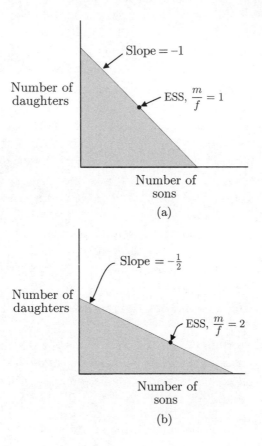

Figure 7.3: When the costs of sons and daughters differ. (a) When mothers have a fixed amount of resources and males and females cost the same, the ESS sex ratio is one to one. (b) When daughters are twice as expensive as sons, the ESS sex ratio is two males for each female.

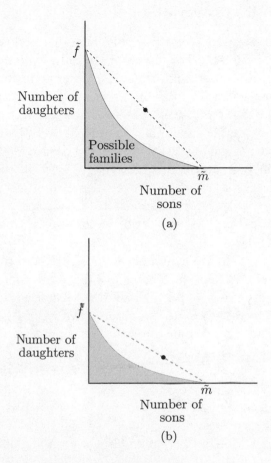

Figure 7.4: When the fitness set is concave, it is cheaper to make families that are either all males or all females. The Shaw-Mohler theorem says that they should be produced in proportion to the relative costs. (a) ESS is one family of daughters per family of sons. (b) ESS is two families of sons per family of daughters.

fitness varies among parents and (2) when population structure causes local competition for mates or resources.

7.4.1   When fitness varies

In some species, maternal condition influences her ability to raise offspring of one sex. For example, in many mammals, offspring of females in good condition grow faster and become bigger than sons of females in poor condition. This benefits sons more than daughters because size matters in male-male competition. In parasitic wasps that lay eggs in caterpillars it pays to lay female eggs in bigger caterpillars because females who grow up in big caterpillars grow larger and lay more eggs. In such cases, selection favors females who adjust the sex ratio of their offspring in response to the situation. In this section, we extend the sex ratio models above to deal with this possibility.

Suppose that there are a number of different situations, numbered by $i \in 1, 2, 3, \ldots, n$. The probability that a female finds herself in situation i is p_i. Sons of females in situation i have reproductive value V_i and daughters have reproductive value U_i. Suppose that females in situation i produce fractions r_i of sons and $(1 - r_i)$ of daughters. We want to find the ESS values of $r_1, r_2, \ldots, r_n$.

It is easy to show that the Shaw-Mohler theorem is equivalent to saying that the ESS sex ratio maximizes the product $f \cdot m$. To determine the ESS sex ratio, we make use of this form of the Shaw-Mohler theorem. Using the assumptions above, we obtain

$$m \cdot f = \underbrace{\left(\sum_i p_i V_i r_i \right)}_{m} \underbrace{\left(\sum_i p_i U_i (1 - r_i) \right)}_{f}.$$

We want to find the fraction of males in each situation that maximizes the product. Taking the derivative with respect to r_i (a particular value of i, such that all other r's are evaluated as constants), we obtain

$$\frac{\partial(m \cdot f)}{\partial r_i} = p_i V_i \left(\sum_i p_i U_i (1 - r_i) \right) - p_i U_i \left(\sum_i p_i V_i r_i \right)$$

$$= p_i \left(V_i f - U_i m \right).$$

Notice that the derivative is not a function of r_i. This means that it is either positive or negative for all values of r_i. If it is positive, making more males yields higher fitness. If it is negative, making more females yields higher fitness. Thus, making more females is favored whenever

$$\frac{U_i}{V_i} > \frac{f}{m}.$$

This says that females in a particular situation (e.g., good condition) should produce all of the sex that has the higher relative reproductive value among the offspring in that situation. It doesn't matter if females in good condition can make better daughters as well as better sons compared to females in poor condition. What matters is whether sons or daughters do better relative to the other. Its like the concept of comparative advantage in the theory of international trade. This will be more intuitive if you consider the following question: should a superstar athelete like David Beckham mow his own lawn, or should he hire somebody to do it? He would undoubtedly be a great lawn mower, fast and stylish, quite likely better than anybody he could hire. But obviously he should hire somebody else to do it if it interferes in any way with his conditioning because he is even better at soccer than at mowing the lawn. It's the same with sex ratio. A female who can make better sons *and* daughters than other females should nevertheless let other females make all the daughters if her own sons are even better than her own daughters.

These ideas were first laid out in a paper by Robert Trivers and the mathematician D. E. Willard, and this result is often called the *Trivers-Willard effect*. Unfortunately the original Trivers-Willard paper[95] is brief and hard to understand. As a result many students think that females in good condition should invest more in the sex with the higher variance in reproductive success (RS). However, it is not the variance that matters, but the effect of increased investment. It is often true in mammals that the sex with greater variance in RS (males) benefits more from investment. But it is possible that the sex with the lower variance in RS benefits more, depending upon the details of the biology.

Olof Leimar[96] produced a model of the Trivers-Willard effect that shows exactly this. The Trivers-Willard argument most people know invokes reproductive success, not reproductive value. Properly incorporating reproductive value into the model, as Leimar did, shows that female-biased sex ratios can be favored if mothers in good condition produce daughters who will be in good condition when they come to reproduce. In many species of monkeys, for example, daughters inherit their mother's dominance rank, and consequently daughters of high-ranking mothers have higher reproductive value than sons, who do not inherit maternal rank. Sons of high-rank mothers may have high reproductive success, but they cannot transmit this value to their own sons. Thus their son's success may be short lived, while the success of a daughter can translate into an enduring investment. To make the right predictions about sex ratio, we need to understand how the life history of a given species generates the reproductive value of each sex. This subtlety may explain why empirical support for the original Trivers-Willard hypothesis is mixed.[97]

7.4.2 Population Structure

As we have seen there are many situations in which population structure can lead to nonrandom social interaction. Here we will see that population structure can lead to a biased sex ratio. This idea, first laid out clearly in a paper by W. D. Hamilton, has provided some of the most beautiful and satisfying empirical tests of evolutionary theory ever devised.

The model is patterned after the biology of fig wasps, small wasps that lay their eggs inside figs. Somewhere between one and a few wasp females lay their eggs inside the fig. The wasps hatch, grow, and pupate in the fig. When they become adults, they mate, still inside the fig. Then the males die (yep, inside the fig), and the females emerge, fly around, find another fig, lay their eggs, and the wheel has turned one more time.

Suppose that n females chosen at random from a large population lay eggs in one of a large number of mating arenas. Individuals mature in the arenas and mate at random with the offspring of all n females, including their own siblings. Males then die, and females disperse at random. Suppose that there are two types of females. A common type produces rb sons and $(1-r)b$ daughters. However, there are a few rare mutant females who divide their reproductive effort between $r' = r + d$ sons and $1 - r' = (1 - r - d)$ daughters, where d is a small number. The mutant type can invade provided

$$W(r|n-1) < W(r'|n-1)$$

where $W(r|n-1)$ and $W(r'|n-1)$ are, respectively, the fitnesses of a common- and a mutant-type female given that the other $n-1$ females in the group are common (r) types. Before you start writing both fitness expressions above and working through the algebra, let us introduce another method that is often less tedious but conceptually more subtle. Since the d in $r' = r + d$ is small, we can ignore terms of order d^2 and use a linear approximation to find the ESS. We did this in each previous CSS solution. This allows us to approximate $W(r'|n-1)$ by

$$W(r'|n-1) \approx W(r|n-1) + d \left. \frac{\partial W(r'|n-1)}{\partial r'} \right|_{r'=r}.$$

This just says that the fitness of the mutant is approximately the fitness of a common female plus the size of the deviation in behavior (d) times the linear effect a change in behavior has on fitness (the partial derivative—this is what those loopy d's

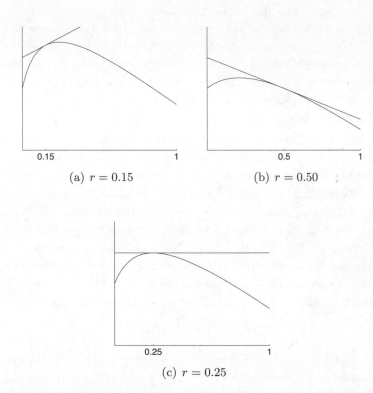

(a) $r = 0.15$ (b) $r = 0.50$

(c) $r = 0.25$

Figure 7.5: Illustration of why the optimal r is found via the derivative method explained in the main text. In these figures, $b = 10$
and $n = 2$. Each plot is $W(r', n - 1)$ as a function of r' for three
different values of the common-type sex ratio r. The straight line
in each plot is the slope of the fitness function at $r' = r$. This is
the linear estimate of the fitness function at the common-type sex
ratio. (a) Values of $r' > r$ lead to increased fitness, as the slope
at this point is positive. Mutants with more sons invade. (b) Now
values of $r' < r$ lead to increased fitness. Mutants with fewer sons
invade. (c) The slope at $r = 0.25$ is zero, meaning that small increases and decreases both lead to lower fitness. This is the ESS
proportion of sons, found where $\left.\frac{\partial W(r'|n-1)}{\partial r'}\right|_{r'=r} = 0.$

indicate) in the region of the common type (hence, evaluated at $r' = r$—this is what the tall bar on the right indicates). Then r' can invade if

$$W(r|n-1) + d \left. \frac{\partial W(r'|n-1)}{\partial r'} \right|_{r'=r} > W(r|n-1)$$

$$d \left. \frac{\partial W(r'|n-1)}{\partial r'} \right|_{r'=r} > 0.$$

Suppose that the partial derivative tells us that increases in r' lead to increases in fitness. Then if $d > 0$, the mutant will invade. Similarly, if $d < 0$, the common type is stable. This means that we only have to write a single fitness function, $W(r'|n-1)$, and its derivative, in order to determine if the mutant can invade. The CSS solution will exist where

$$\left. \frac{\partial W(r'|n-1)}{\partial r'} \right|_{r'=r} = 0.$$

We provide a graphical explanation in Figure 7.5.

Remember, this method is equivalent to comparing the two fitness expressions, assuming that the deviation in behavior d is small so that terms of order $d^2 \approx 0$, as we did in previous chapters. However, doing the calculus here is often a lot less tedious than expanding and factoring polynomials. If you are not comfortable with derivatives, you may not agree.

In Box 7.4, we show you how to use this tactic to deduce that the evolutionarily stable sex ratio is

$$r^\star = \frac{n-1}{2n}. \tag{7.4}$$

Alright, but what does this mean? As the following table shows, when n is large, the ESS sex ratio is approximately one-half, the same as the Fisherian result. But for small values of n, sex ratios are strongly biased toward females. When $n = 1$, the model says females should produce only daughters, which is obviously impossible. What this really means is that females should produce only as many males as they need to fertilize all their daughters—the dairy farmer optimum.

Box 7.4 Deriving the ESS sex ratio under local mate competition

Let's first write the expression for the fitness of the mutant female with sex ratio $r' = r + d$. Recall that b is the family (brood) size and n is the number of females who lay in a single fig. As before, v_f and v_m are the reproductive values of females and males. Then total fitness of a mutant female in a group with $n-1$ common-type females is

$$W(r'|n - 1) = \underbrace{b(1 - r') \cdot v_f}_{\text{fitness through females}} + \underbrace{br' \cdot v_m}_{\text{fitness through males}}$$

As in the Fisherian model, $v_m n_m = v_f n_f$, but now the numbers of males and females that matter are only those in the local group (fig). This implies

$$n_f = b(1 - r') + b(n - 1)(1 - r)$$
$$n_m = br' + b(n - 1)r.$$

Let $v_f = 1$, making $v_m = n_f/n_m$. (Remember, reproductive value is relative.) Incorporating the reproductive values gives

$$W(r'|n - 1) = b(1 - r')(1) + br' \left(\frac{b(1 - r') + b(n - 1)(1 - r)}{br' + b(n - 1)r} \right).$$

Taking the derivative with respect to r' gives

$$\frac{\partial W(r'|n - 1)}{\partial r'} = -b + \left(\frac{r' + (n - 1)r - r'}{(r' + (n - 1)r)^2} \right) \{b(1 - r')$$
$$+ b(n - 1)(1 - r)\} + \left(\frac{r'}{r' + (n - 1)r} \right)(-b).$$

To find the ESS value of r, $r^\star$, set $r' = r$ and simplify. Then set the expression equal to zero and solve for r:

$$\left. \frac{\partial W(r'|n - 1)}{\partial r'} \right|_{r'=r} = b \left(-1 + (1 - r)n \left(\frac{(n - 1)r}{n^2 r^2} \right) - \frac{1}{n} \right) = 0$$
$$-nr + (1 - r)(n - 1) - r = 0.$$

Solving this for r yields the solution, Expression 7.4 in the main text.

n	1	2	4	8	100
$r^\star$	0	$\dfrac{1}{4}$	$\dfrac{3}{8}$	$\dfrac{7}{16}$	$\dfrac{99}{200}$

So what is going on here? The answer is that selection between groups favors female-biased sex ratios, while selection within groups favors even sex ratios. The only way for a group to reproduce is through females. Thus selection between groups favors a female-biased ratio. As groups get smaller the amount of variation between groups increases, and sex ratio becomes more biased. As groups get larger, they come to be better and better samples of the global population—variance among groups declines—and the model results in behavior closer and closer to the Fisherian result.

7.5   Breaking the eigen barrier

In Chapter 1, we said that the essence of Darwinism is to create an accounting system that keeps track of the processes that change gene frequencies, and then iterates these processes over generations to determine the long-run evolutionary outcome. Reproductive value is a clever trick that allows us to dodge the iterating part, but why *can't* we do it the old-fashioned way? Write down recursions, find the equilibria, and determine their stability?

The answer is we can, but doing it requires a new mathematical trick. Here's why: a gene that distorts sex ratio acts in females (changing the number of sons and daughters they have) but is also carried in males. The frequency of these genes in males and females might—and probably will—differ because a gene that creates more daughters will be over-represented in daughters. So we must keep track of two-state variables, the frequency of sex ratio distorters in females, and the frequency in males. This means we have a model with two state variables and two coupled recursions, one for males and one for females. There is no problem finding the equilibrium values—this is done just as you'd expect.

However, we can't use the method presented in Chapter 1 to determine their stability. We have to use a method based on matrix algebra, and this method is often intimidating for novices—so much so that a graduate student friend of ours used to call this the "eigen barrier." The problem is that to really master these matrix methods you have to learn a bunch of new mathematics, well beyond the usual introductory calculus course.

There are a number of good reasons to learn these methods. First, population biologists use them all the time, so you will read lots of papers which assume that you know about "matrix equations" and "eigen values" and the like. You need to understand something of these methods to make sense of much of the literature. Second, there are problems that are actually easier to solve in matrix formulation because people have already studied analogous systems and written exact solutions in the form of matrix equations. If you can translate your system into a matrix equation, then you can benefit from their hard work. Third, while you can duck matrix equations and eigen values in the case of sex ratio by using reproductive value, reproductive value calculations usually depend on very similar methods. In this section, we introduce these methods, using the example of Fisherian sex ratio. If you are like most people, you will need to read about these methods a few times and practice them yourself before you will really grasp their use.

Consider a large diploid population. At a particular locus the A allele is common and the a allele is rare. A female's genotype at this locus affects the mix of males and females she produces. AA females produce f females and m males, while Aa females produce $f^\star$ females and $m^\star$ males. We want to know whether the a allele can invade when rare, so we don't care about the phenotype of aa females because they will always be very rare. A male's genotype at this locus has no effect on his phenotype.

Because the genes at this locus affect the numbers of sons and daughters that a female has, the frequency in males and

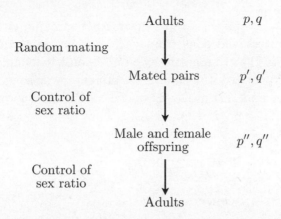

Figure 7.6: The life cycle assumed in the model of the evolution of sex ratio. Adults with frequencies of the a allele p and q for females and males, respectively, mate randomly. Mated pairs produce offspring, the composition of which is determined by the female's genotype. Another phase of sex ratio control can occur after birth as offspring mature into adults.

females won't necessarily be the same. Suppose, for example, that Aa females make more daughters. Then the frequency of the a allele can be higher in females. This means we have to keep track of the frequency of different types in the two sexes separately. Let $p = \text{freq}(Aa \text{ females})$ and $q = \text{freq}(Aa \text{ males})$ among adults just before mating takes place. Since a is rare, we assume that p and q are small numbers. Next, assume the life cycle shown in Figure 7.6.

We want to keep track of how the frequencies p and q change through time. Random mating results in the following mating table (refer back to Chapter 1 if you've forgotten about mating tables):

Mating			Daughters		Sons	
Female	Male	Probability of mating	AA	Aa	AA	Aa
AA	AA	$(1-p)(1-q) \approx 1-p-q$	f	0	m	0
AA	Aa	$(1-p)q \approx q$	$\frac{1}{2}f$	$\frac{1}{2}f$	$\frac{1}{2}m$	$\frac{1}{2}m$
Aa	AA	$p(1-q) \approx p$	$\frac{1}{2}f^{\star}$	$\frac{1}{2}f^{\star}$	$\frac{1}{2}m^{\star}$	$\frac{1}{2}m^{\star}$

To get the frequencies after mating we sum the numbers of Aa daughters and sons from each mating weighted by the probability that the mating occurs. Then we divide this sum by the expected total number of daughters or sons (of both types–AA and Aa) in the next generation. So the frequency of Aa females, p', is

$$p' = \frac{q(\frac{1}{2}f) + p(\frac{1}{2}f^\star)}{(1 - p - q)f + qf + pf^\star}. \tag{7.5}$$

Similarly, the frequency of Aa males is

$$q' = \frac{q(\frac{1}{2}m) + p(\frac{1}{2}m^\star)}{(1 - p - q)m + qm + pm^\star}. \tag{7.6}$$

Next, suppose that there is sex specific mortality. For example, in many species males are more likely to die before reaching adulthood than are females. If, as is likely the case, a is more common among females, sex-specific mortality will definitely change the frequency of a in the population as a whole. However, it will not change the frequency among females because females have the same mortality on average. Thus the frequencies among adults in the next generation are also p' and q'.

This is a new mathematical object for us, a two-dimensional set of recursions with the form

$$p' = F(p, q)$$
$$q' = G(p, q).$$

Dynamic equations like the ones we use in evolutionary theory are based on a set of variables that are sufficient to summarize everything of interest about the system. These variables are called *state variables*. So far we have dealt mainly with models with only one state variable (when there were two we just waved our arms, rapidly). Now we want to introduce the tools used to analyze models with two or more state variables.

First, we look for equilibrium values that satisfy the equations

$$\hat{p} = F(\hat{p}, \hat{q})$$
$$\hat{q} = G(\hat{p}, \hat{q}).$$

That is, we want to find the state or states of the system that lead to no change.

In the present system we are interested in whether a population in which A is common can resist invasion by a. In other words, we want to know whether $p = 0$ and $q = 0$ is a stable equilibrium. By substituting these values into the recursions 7.5 and 7.6 derived above, we confirm that $(0, 0)$ is an equilibrium. Next we want to know if the equilibrium is stable. As in the one-dimensional case, we linearize the nonlinear recursion in the neighborhood of the equilibrium point. So

$$p' \approx p \left. \frac{\partial F}{\partial p} \right|_{\hat{p}, \hat{q}=0} + q \left. \frac{\partial F}{\partial q} \right|_{\hat{p}, \hat{q}=0}$$

$$q' \approx p \left. \frac{\partial G}{\partial p} \right|_{\hat{p}, \hat{q}=0} + q \left. \frac{\partial G}{\partial q} \right|_{\hat{p}, \hat{q}=0}.$$

This has the same logic as linearization in the one-dimensional case. For each state variable, we approximate the true nonlinear recursion by a linear equation (in this case a plane) at the equilibrium point. As long as we are close to the equilibrium point this will be adequate. The partial derivatives mean just take the derivative of the function with respect to that variable, treating the others like constants. So, for example,

$$\frac{\partial F}{\partial p} = \frac{(\frac{1}{2}f^\star)\left(((1-p-q)f+fq+f^\star p)-(-f+\frac{1}{2}f^\star)(\frac{1}{2}fq+\frac{1}{2}f^\star p)\right)}{((1-p-q)f+qf+pf^\star)^2}.$$

Substituting in the equilibrium values $\hat{p} = 0$ and $\hat{q} = 0$ yields

$$\left. \frac{\partial F}{\partial p} \right|_{\hat{p}, \hat{q}=0} = \frac{\frac{1}{2}f^\star}{f}.$$

Taking the other three partial derivatives and substituting the resulting values yields the following linearized recursions in the neighborhood of the $(0, 0)$ equilibrium:

$$p' \approx p\frac{\frac{1}{2}f^\star}{f} + q\frac{1}{2} \qquad (7.7)$$

$$q' \approx p\frac{\frac{1}{2}m^\star}{m} + q\frac{1}{2}.$$

It is useful to rewrite these using matrix algebra as follows:

$$\begin{pmatrix} p' \\ q' \end{pmatrix} = \frac{1}{2}\begin{pmatrix} \frac{f^\star}{f} & 1 \\ \frac{m^\star}{m} & 1 \end{pmatrix}\begin{pmatrix} p \\ q \end{pmatrix}.$$

This is just the matrix form of the two recursions in (7.7).

It turns out that multidimensional linear recursions like this one have the same general property as one-dimensional recursions, namely they either grow away from the equilibrium or shrink back toward it. Which happens depends on what is called the *dominant eigen value* of the matrix that defines the recursion (above, this matrix is the 2×2 matrix). So what is an eigen value, anyhow?[98] Suppose $\mathbf{A}$ is an $n \times n$ matrix and $\mathbf{x}$ is a $1 \times n$ column vector. Then the eigen values λ of the matrix $\mathbf{A}$ satisfy the matrix equation

$$\mathbf{Ax} = \lambda\mathbf{x}.$$

What we are doing here is solving for the equilibrium growth rate λ of each element in the system defined in $\mathbf{x}$. For most dynamic systems of this type, the system eventually reaches a steady state wherein all state variables grow (or shrink) at the same constant rate. The matrix equation above simply states this result. In more familiar notation,

$$p \cdot a_{11} + q \cdot a_{12} = \lambda p,$$
$$p \cdot a_{21} + q \cdot a_{22} = \lambda q.$$

So perhaps now you can see that what we are asking is whether there are any values λ that let us simultaneously update both (all) state variables, as if we implemented the recursions defined in the matrix $\mathbf{A}$, with elements a_{ij}.

It turns out that for most matrices $\mathbf{A}$ of size $n \times n$, there are n eigenvalues, each a solution to the system of equations. These things are extremely useful in all kinds of mathematics. Here they are interesting because it turns out that the behavior of the linearized recursion depends on the magnitude of the largest eigen value. If we can find the largest growth rate λ, we can see if the system will grow or shrink away from the equilibrium. If it is greater than 1, a perturbed system grows away from the equilibrium. If it is less than 1, the system shrinks back to the equilibrium.

So how do you calculate eigen values? The answer for small (2×2 and 3×3) matrices is pretty easy, but for larger matrices it is possible only with special tricks that are beyond the scope of this book. Here's how you do it for small matrices. You can rearrange the matrix equation above to be,

$$\mathbf{Ax} - \lambda\mathbf{x} = 0$$

$$(\mathbf{A} - \lambda\mathbf{I})\mathbf{x} = 0, \tag{7.8}$$

where $\mathbf{I}$ is the "identity matrix," a matrix with 1's down the diagonal and 0's everywhere else. If you multiply any matrix of the same dimension by $\mathbf{I}$, you get that matrix back unchanged. So it's the matrix version of the integer 1. We had to perform this step because λ was, implicitly, factored out of the matrix $\mathbf{x}$. So we couldn't begin by dividing both sides by $\mathbf{x}$. Now that we have a new matrix, $\mathbf{I}$, to multiply by λ, all is well.

Now, Expression 7.8 represents a set of *homogeneous* linear equations. You learned in high school that such equations only have a solution if the *determinant*[99] of the matrix is zero. That is,

$$\det(\mathbf{A} - \lambda\mathbf{I}) = 0.$$

In our case,

$$(\mathbf{A} - \lambda \mathbf{I}) = \begin{pmatrix} \frac{1}{2}\frac{f^\star}{f} & \frac{1}{2} \\ \frac{1}{2}\frac{m^\star}{m} & \frac{1}{2} \end{pmatrix} - \begin{pmatrix} \lambda & 0 \\ 0 & \lambda \end{pmatrix} = \begin{pmatrix} \frac{\frac{1}{2}f^\star}{f} - \lambda & \frac{1}{2} \\ \frac{\frac{1}{2}m^\star}{m} & \frac{1}{2} - \lambda \end{pmatrix}.$$

The determinant of an $n \times n$ matrix is an nth-order polynomial. Thus, finding the eigen values of a matrix is the same as finding the roots of an nth-order polynomial. Easy if $n = 2$, tricky if $n = 3$, and you have to be lucky for $n > 3$. Fortunately $n = 2$ in our case, so let's forge ahead. The determinant is

$$\det(\mathbf{A} - \lambda \mathbf{I}) = \left(\frac{\frac{1}{2}f^\star}{f} - \lambda \right) \left(\frac{1}{2} - \lambda \right) - \frac{1}{2} \left(\frac{\frac{1}{2}m^\star}{m} \right)$$

$$= \lambda^2 - \frac{1}{2}\lambda \left(1 + \frac{f^\star}{f} \right) + \frac{1}{4} \left(\frac{f^\star}{f} - \frac{m^\star}{m} \right). \quad (7.9)$$

Now, Equation 7.9 is quadratic, so there are two values of λ that are solutions. The bigger of these is the dominant eigen value, which we will label $\lambda^\star$. If it is less than 1, the equilibrium is stable. We use the quadratic formula[100] to solve for λ:

$$\lambda = \frac{\frac{1}{2}\left(1 + \frac{f^\star}{f}\right) \pm \sqrt{\frac{1}{4}\left(1 + \frac{f^\star}{f}\right)^2 - 4 \cdot \frac{1}{4} \cdot \left(\frac{f^\star}{f} - \frac{m^\star}{m}\right)}}{2}.$$

It will simplify the algebra if we make the following substitutions: $f^\star = f + \phi$ and $m^\star = m + \mu$. The new parameters ϕ and μ represent the change in sex ratio caused by the mutant allele. These substitutions give us

$$\lambda = \frac{\left(1 + \frac{\phi}{2f}\right) \pm \sqrt{\left(1 + \frac{\phi}{2f}\right)^2 - \left(\frac{\phi}{f} - \frac{\mu}{m}\right)}}{2}.$$

The common-type sex-ratio gene AA, which leads to families with f daughters and m sons, can resist invasion by the mutant if the biggest of the two eigen values specified above is less than one($\lambda^\star < 1$). Since the square root term is positive, the dominant eigen value will be the "+" root, and thus stability requires

$$1 > \lambda^\star$$

$$2 > \left(1 + \frac{\phi}{2f}\right) + \sqrt{\left(1 + \frac{\phi}{2f}\right)^2 - \left(\frac{\phi}{f} - \frac{\mu}{m}\right)}$$

$$\left(\frac{\phi}{f} + \frac{\mu}{m}\right) < 0.$$

This is the same as Expression 7.3 but with the inequality reversed. Thus the mutant invades if the inequality does not hold. The same method can be used to determine stability in many dynamic systems, in which state variables represent geographic structure, age structure, or any other aspect of life history that affects reproductive value.

Guide to the Literature

Reproductive Value. For deeper understanding of reproductive value, consult Caswell's 2001 patient textbook, *Matrix Population Models*. It provides several useful interpretations of the concept and demonstrates how to derive these values from assumptions about the life history of an organism. Another approach that uses the same concepts, but applied to problems more like those we consider here, is given by Houston and McNamara 1999. **Fisherian sex ratio.** Fisher's 1930 book presented a verbal argument for natural selection equalizing parental expenditures on each sex. Charnov's 1982 book explores many aspects of the Shaw-Mohler theorem and alternative life histories on sex allocation. Stephen Frank's 1998 book contains a useful, if somewhat technical, explanation of different theoretical factors affecting sex ratio. Pen and Weissing's 2002 chapter on sex allocation theory contains a stimulating "model gallery," as well as a patient explanation of how reproductive value factors in sex ratio models. **Sex**

ratio adjustment. The original Trivers and Willard 1973 paper continues to generate interest. Evidence of the effect is strong in some taxa, such as wasps, but generally weak in mammals. See Brown and Silk 2002, West and Sheldon 2002, and chapters in Hardy's 2000 *Sex Ratios: Concepts and Research Methods*. Olof Leimar's 1996 derivation of the theory is essential reading for empirical research and theorists alike. **Local mate competition and sex allocation.** Our presentation is based on Hamilton's 1967 derivation. See also models by Michael Bulmer and Peter Taylor (Bulmer and Taylor 1980, Taylor and Bulmer 1980). Aviles 1993 provides another example of how population structure can affect sex ratio. The issue is closely related to viscosity, as discussed in Chapter 3. **Multidimensional stability analysis.** We recommend Alan Hastings' 1997 introductory text, *Population Biology: Concepts and Models*, as well as the very thorough textbook by Caswell 2001. McNamara and Houston 1996 proved the equivalence of linear stability methods and those based on reproductive value. See also Taylor 1996, who spells this argument out beginning on page 663.

Problems

1. Big caterpillars and female wasps. Female wasps live in an environment in which there are two kinds of caterpillars, big and little. The probability of finding a big caterpillar is g and the probability of finding a little caterpillar is $1 - g$. The females lay k eggs on each caterpillar. The fitness of female eggs laid on a big caterpillar is two times the fitness of female eggs laid on a small caterpillar. Male eggs have the same fitness on either type of caterpillar. (a) Suppose that the strategy of producing all females on big caterpillars and all males on small caterpillars is common in a population. Show that this population can resist invasion by genes that lead to the production of some male eggs on big caterpillars whenever $g < 1/2$. (b) Show that the population can resist invasion by genes leading to the production of some females on small caterpillars whenever $g > 1/3$. (c) Suppose that g = 1/4. This means that genes that lead to the production of a mix of male and female eggs on small caterpillars can invade a population in which individuals lay only male eggs on small caterpillars. Show that genes that reduce the fraction of males produced on small caterpillars will continue to invade until the proportion of males produced on small caterpillars is $r* = 5/6$. Also show that producing all females on large caterpillars continues to be an ESS.

2. Inbreeding and local mate competition. Richard Alexander (Alexander and Sherman 1977) argued that local mate competition favors female-biased sex ratios because there is brother sister mating—selection among groups has nothing to do with it. Modify the local mate competition model described in the text by assuming that brothers and sisters can recognize each other and thereby avoid mating with each other. Show that the ESS proportion of males in this model is

$$r^\star = \frac{n-2}{2n-3},$$

where n is the number of females who lay eggs on a single island/fig. How do you interpret this result?

3. Cooperation among siblings. In many species, offspring of one sex remain in their natal group and cooperate for their mutual benefit, while the offspring of the other sex emigrate to a variety of different groups and thus are not able to cooperate. For example, in most human hunting and gathering societies males usually remain home and cooperate in hunting and group defense, and in most monkey species, females remain and cooperate with each other in both intra- and intergroup competition. In this problem you will analyze a simple model which shows that such sex-biased cooperation can lead to the evolution of a biased sex ratio. Suppose that in a hypothetical species, females give birth to R offspring, m males and f females. Assume that males and females require equal investment so that

$$R = m + f.$$

After parental care is terminated but before they reproduce, both sexes experience mortality. Males remain in their natal group and cooperate, while females disperse. Because related males cooperate, the probability of male survival to adulthood in a family with m sons and f daughters is m/R. The probability that females survive, a, does not depend on family composition. Thus,

$$m' = \frac{m^2}{R}$$

and

$$f' = af.$$

Prove that ESS fraction of males at birth, m/R, is $\frac{2}{3}$.

4. More than one constraint. In a hypothetical species, family size and composition are limited by two different resources—energy and water. Male offspring require twice as much energy as female offspring, and the amount of energy available would allow mothers to produce E daughters or $\frac{E}{2}$ sons. Female offspring require three times as much water as do males, and there is enough water to allow females to produce W sons or $\frac{W}{3}$ daughters. Careful studies suggest that $E \approx W$.

(a) Draw the fitness set, and label the intersection points with the f and m axes and any other salient features.

(b) Use the Shaw-Mohler theorem to determine the ESS sex ratio.

Notes

[88]Not to be confused with W. D. Hamilton of inclusive fitness fame.

[89]See page 142 of the 1958 edition.

[90]Fisher actually invoked reproductive *value*, which we will get to later.

[91]Fisher defined reproductive value as follows: "To what extent will persons of this age [or sex], on average, contribute to the ancestry of future generations." We find the passage to be quite confusing. We don't recommend it until you already understand the concept.

[92]Taylor's 1996 paper shows how to deduce this from a gene flow matrix defined by the reproductive biology of a species.

[93]Shaw and Mohler 1953.

[94]See Pen and Weissing 2002.

[95]Trivers and Willard 1973.

[96]Leimar 1996.

[97]See for discussion Brown and Silk 2002 and West and Sheldon 2002.

[98]The term comes from German, where it is *eigenwert*, which can be glossed as "characteristic value." These values are characteristic of the matrix we compute them from. For some odd reason, we retain the German *eigen* in English, rather than call them simply the characteristic values of the matrix.

[99]Remember, if $\mathbf{A} = \left(\begin{smallmatrix} a & b \\ c & d \end{smallmatrix} \right)$, then $\det \mathbf{A} = ad - cb$.

[100]For those who have forgotten, the solutions to the equation $0 = ax^2 + bx + c$ are given by

$$x = \frac{-b \pm \sqrt{b^2 - 4ac}}{2a}.$$

Box 7.5 Symbols used in Chapter 7

f Number of daughters produced by a common female

m Number of sons produced by a common female

R Average family size (number of grandkids)

$f^\star, m^\star$ Daughters and sons, respectively, produced by a mutant female

d Number of additional daughters and fewer sons produced by a mutant female

λ Eigen value of a projection matrix, long-term population growth rate

v_f, v_m Reproductive value of females and males, respectively

n_f, n_m Numbers of females and males, respectively

r_i For a female in state i, fraction of offspring which are sons

U_i, V_i Reproductive values of sons and daughters, respectively, of a female in state i

p_i Probability female is in state i

r Common sex ratio of offspring, sons/daughters

r' Mutant sex ratio of offspring

n Number of females in each patch

b Average family (brood) size of each female

Chapter 8

Sexual Selection

Darwin's second-most-famous book is usually cited simply as *The Descent of Man*, but its real title is *The Descent of Man and Selection in Relation to Sex*, and the second half of the treatise is devoted to what we now call sexual selection— selection on individuals of the nonlimiting sex (usually males) caused by differential access to the limiting sex (usually females). Darwin's main concern was to explain the evolution of traits in males such as the peacock's tail that seem grossly maladaptive in conventional terms.

Following Julian Huxley,[101] most people divide sexual selection into two components. *Intrasexual* selection refers to selection caused by male-male competition. For example, in species like rhesus macaques, males fight each other over access to females. As a consequence, selection favors large body size or large canines. *Intersexual* selection refers to selection caused by female preferences. For example, peahens preferentially mate with peacocks that have many eye spots in their tails. This can create strong selection favoring males with bigger tails with more eyes spots.

If you think about this distinction very hard, it gets very fuzzy. Supposed examples of intrasexual selection always depend on female (or male) preferences, if, as seems sensible, you define preferences as any female (or male) characteristic that affects whom females (or males) mate with. For example, the behavior of female rhesus macaques can affect whom they mate with. If this behavior were different, so that they refused to mate with the winners of male dominance contests, then selection would work differently on males. If you don't think that this is plausible, think again. Spotted hyenas live in a multimale social structure very much like macaques, and there is substantial variation in male mating success. However, male spotted hyenas are smaller than females. The same is true of examples of intersexual selection. Although female preferences drive mate choice, clearly, peacocks are competing to attract females, and such competition can be spectacularly evident on leks. In every case in which males compete, the nature of that competition is affected by female preferences.[102]

It seems to us that the right way to distinguish these examples is to ask: What shapes the evolution of female (male) preferences? Is it mainly something other than the identity of the male (female) they mate with? This is probably the case in competition among male macaques. Or is the evolution of female preferences mainly driven by the consequences of these preferences for mate choice? The second case is often considered the most interesting, and in this chapter we will consider models of the evolution of female preferences for different male characters. Like Darwin, we will be mainly interested in the evolution of preferences for seemingly maladaptive, exaggerated male characters.

Much of the literature divides explanations for the evolution of female preferences into three distinct categories:

1. Fisherian runaway models in which preferences evolve as a correlated response to the selection on male traits that these preferences induce.

2. Good-genes models in which males signal the qualities of their genes by developing traits that serve as costly signals of those qualities.

3. Sensory bias models in which female preferences for exaggerated male characters are side effects of an adaptive sensory system.

These distinctions are useful, but we will see that runaway and good-genes models are fundamentally similar. They are both about the costs and benefits of being choosy and whether females can increase the fitness of their daughters or sons by being choosy.

This chapter also presents an opportunity to explain a new set of modeling tools. Models of the evolution of female preferences require modeling the evolution of at least two characters—the female preference and the male trait. Several different approaches have been adopted. Some work has been based on "major-gene" models in which each trait is controlled by a single genetic locus. Thus this approach generates multi-locus genetic models. More work has been based on quantitative genetic models in which each trait is controlled by a large number of loci, and it is assumed that the distribution of additive genetic values is distributed normally. This approach generates models that keep track of the evolution of the mean values of each trait. Finally, several authors have recently developed game-theoretic models that also involve multiple traits but are simpler to analyze.

In this chapter we begin by focusing on the quantitative genetic models. We will study Russ Lande's influential model of the runaway process as well as its descendants. These models help us understand why small costs and benefits for females can have big effects on males. We will see how these results make sensory bias models plausible. Then we turn to a recent game-theoretic model by Hanna Kokko, John McNamara, and their coworkers that provides a very clear understanding of the relationship between Fisherian models and so-called good-genes models. This model is the first in the book that must be solved numerically, by assuming values for all the parameters and finding ESS solutions with a computer.

8.1 Quantitative genetic models

Quantitative genetic models assume that observable phenotypic traits are affected by many genes at many different genetic loci, and, as a consequence, that there is a continuum of possible genetic values. For example, a trait like height in humans may take on any value over a wide range, not a set of discrete values.

There are a number of different approaches to modeling the evolution of quantitative characters. We have adopted the approach used by Yoh Iwasa and Andrew Pomiankowski, which is based on the assumption of weak selection.[103]

8.1.1 The evolution of a single quantitative character

To keep things simple, we will begin with a single quantitative trait with complete heritability. The trait is some aspect of phenotype that can be measured, like body size or canine length. By completely heritable, we mean that all the differences between individuals in this trait are due to genetic differences—there is no environmental variation. Let x be the value of the character in an individual, and let $n(x)$ be the number of individuals with trait value x. The population size is

$$N = \sum_x n(x),$$

where N is very large. Then, the mean and variance of the trait in the population are, respectively,

$$\bar{x} = \frac{1}{N} \sum_x x n(x)$$

and

$$\text{var}(x) = G_x = \frac{1}{N} \sum_x (x - \bar{x})^2 n(x).$$

We assume that the population undergoes viability selection, and that the probability of survival of an individual with phenotype x is $W(x)$. An individual's survival depends on both its own phenotype and the mean phenotype in the population (this is just as it was in Chapter 1). Then, the mean value of the trait after selection, $\bar{x}'$, is

$$\bar{x}' = \frac{\sum xn(x)W(x)}{\sum n(x)W(x)} = \frac{\sum xn(x)W(x)}{\bar{W}}.$$

The change in the mean as a result of selection is

$$\Delta \bar{x} = \bar{x}' - \bar{x} = \frac{\sum xn(x)W(x) - \bar{x}\bar{W}}{\bar{W}}$$

$$= \frac{\text{cov}(x, W(x))}{\bar{W}}. \tag{8.1}$$

The observant reader will notice the similarity between Equation 8.1 and the Price equation we derived first in Chapter 3 and again in Chapter 6.

In Box 8.1, we explain how the covariance above implies that the change in the mean value of the trait in the population is

$$\bar{x}' - \bar{x} = \Delta \bar{x} \approx G_x \left. \frac{\partial \ln W}{\partial x} \right|_{\bar{x}}, \tag{8.2}$$

where the first term, G_x, is the variance in the trait in the population, and the second, $\left. \frac{\partial \ln W}{\partial x} \right|_{\bar{x}}$, often called the *selection gradient*, gives the *percent* change in fitness caused by a change in phenotypic value. For example, consider the Gaussian function

$$W(x) = e^{\frac{-(x-\theta)^2}{2S}}. \tag{8.3}$$

This fitness function represents stabilizing selection (see Figure 8.1). The probability of survival is maximized for $x = \theta$. The farther x is away from θ, the lower is the chance of survival. The size of the deviation from the optimum is measured in units of S, so big values of S mean that x has to be a long way from θ before survival drops much, and this means that selection is weak. Applying Equation 8.3 to 8.2 yields

$$\Delta \bar{x} \approx G_x \frac{\partial}{\partial x} \left(-\frac{(x-\theta)^2}{2S} \right)_{\bar{x}} = \frac{G_x}{S}(\theta - \bar{x}). \tag{8.4}$$

Box 8.1 The weak selection approximation

For values of x close to $\bar{x}$, the fitness function $W(x)$ can be approximated by using a Taylor series approximation. Here it is, keeping terms of the order $(x - \bar{x})^2$:

$$W(x) \approx W(\bar{x}) + (x - \bar{x}) \left.\frac{\partial W}{\partial x}\right|_{\bar{x}} + \frac{1}{2}(x - \bar{x})^2 \left.\frac{\partial^2 W}{\partial x^2}\right|_{\bar{x}}.$$

The vertical bar with $\bar{x}$ on the right means the partial derivative evaluated at $x = \bar{x}$. If the variance of x in the population is small, most individuals have phenotypic values close to $\bar{x}$, and then we can substitute the above for $W(x)$ in the covariance expression,

$$\text{cov}\,(x, W(x)) = \text{E}\left\{(x - \bar{x})W(x)\right\}$$

$$\approx \text{E}\left\{(x - \bar{x})\left(W(\bar{x}) + (x - \bar{x}) \left.\tfrac{\partial W}{\partial x}\right|_{\bar{x}} + \tfrac{1}{2}(x - \bar{x})^2 \left.\tfrac{\partial^2 W}{\partial x^2}\right|_{\bar{x}}\right)\right\}.$$

Distributing the first $(x - \bar{x})$ into the parentheses and canceling any terms of order $(x - \bar{x})^3$ or higher, we obtain

$$\text{cov}(x, W(x)) \approx \text{E}\left\{(x - \bar{x})W(\bar{x})\right\} + \text{E}\left\{(x - \bar{x})^2 \left.\frac{\partial W}{\partial x}\right|_{\bar{x}}\right\}$$

$$\approx G_x \left.\frac{\partial W}{\partial x}\right|_{\bar{x}}.$$

The simplification above follows because $\text{E}(x - \bar{x}) = 0$ and $\text{E}\left\{(x - \bar{x})^2\right\} = \text{var}(x)$. To a same level of approximation, average fitness is $W(\bar{x})$. Thus we have

$$\bar{x}' - \bar{x} \approx \frac{G_x}{W(\bar{x})} \left.\frac{\partial W}{\partial x}\right|_{\bar{x}} = G_x \left.\frac{\partial \ln W}{\partial x}\right|_{\bar{x}}.$$

The last step is justified because the derivative of the natural log (ln) of a function $F(x)$ is

$$\frac{\partial \ln F(x)}{\partial x} = \frac{1}{F(x)} \frac{\partial F(x)}{\partial x}.$$

Applying this in reverse, and noticing that the derivative is evaluated at $x = \bar{x}$, gives the final expression.

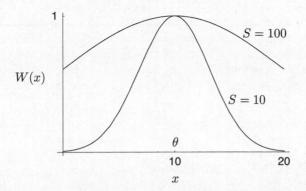

Figure 8.1: Fitness $W(x) = \exp(\frac{-(x-\theta)^2}{2S})$ when $\theta = 10$ for two values of S. This is stabilizing selection because values of x on either side of θ have lower fitness than $x = \theta$. How much lower is determined by S. When $S = 10$, fitness declines rapidly on both sides of the optimal phenotype. When $S = 100$, values of x close to θ have nearly the same fitness as the optimal phenotype.

Thus, if $\bar{x} < \theta$, the mean value of the character increases; if it is greater than the optimum, it decreases. The rate at which this occurs depends on the amount of variation and the strength of selection. In Box 8.2, we discuss how to know whether this approximation is a good one.

Equation 8.4 gives us the effect of selection on the mean. To predict the mean in the next generation we also have to account for the effects of genetic transmission on the mean and the effects of both selection and transmission on the variance. Accounting for the effects of selection on the variance is straightforward. Such models produce two simple and important results:

1. Directional or stabilizing selection reduces the variance.

2. The change in variance is independent of the mean.

Accounting for the effect of transmission requires an explicit genetic model of how many loci are involved, how genotype maps onto phenotype, and the nature of linkage among loci.

Box 8.2 How good is the weak selection approximation?

How good is the approximation given in Equation 8.4? One check comes from an alternative theory that assumes that the trait is normally distributed with mean $\bar{x}$ and variance G_x but makes no assumption about the strength of selection. According to this theory,

$$\Delta\bar{x} = \frac{G_x}{G_x + S}(\theta - \bar{x}).$$

Thus as long as G_x is much smaller than S (and therefore selection is weak), the approximation is okay (see Figure 8.2). The theory that assumes normal distributions is much older, but it is also only an approximation. If the effects of different loci on the trait were additive and independent, the central limit theorem says that the distribution of the trait should be normal. However, Barton and Turelli[104] showed that linkage disequilibrium causes gene effects at different loci to be correlated, and, as a result, the central limit theorem applies only as an approximation. The theory based on normal distributions also does not apply to cases of frequency-dependent selection, while the weak selection approach does.

A variety of models have been analyzed.[105] Many models predict that transmission itself will have no effect on the mean. However, despite working through horrendously complicated mathematics, predicting the evolution of the variance has proven to be intractable, even with heroic simplifying assumptions. Most modelers interested in the evolution of phenotype simply assume that the variance evolves to its equilibrium value independent of changes in the mean and is constant thereafter, and that is what we will assume.

8.1.2   Multiple characters

The final step before we actually turn to sexual selection is to extend this theory to deal with the evolution of more than one character. Suppose that individuals are characterized by two traits. Let $W(x, y)$ be the fitness of an indivdual with trait values x and y. Then a derivation exactly parallel to

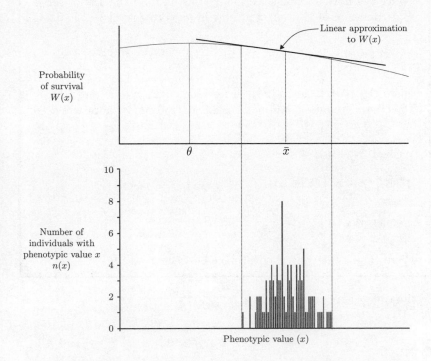

Figure 8.2: The quantitative genetics model with weak selection. The horizontal axis gives phenotypic value x. The lower panel plots the distribution of phenotypic values $n(x)$ in a population of 100 individuals, and the upper panel gives the probability of survival $W(x)$ as a function of phenotypic value x. The fitness-maximizing value of x is θ, so in the situation shown, the mean value of x, $\bar{x}$, is larger than the optimum, and natural selection will cause $\bar{x}$ to decrease. Selection is weak because the fitness function $W(x)$ is much wider than the distribution of phenotypic values. This means that a linear approximation to the fitness function is quite good within the range of variation present in the population.

Box 8.3 Weak selection approximation for two characters

Once again we assume selection is weak and use the Taylor series approximation of $W(x, y)$, but now in two dimensions:

$$W(x, y) \approx W(\bar{x}, \bar{y}) + (x - \bar{x}) \left. \frac{\partial W}{\partial x} \right|_{\bar{x}, \bar{y}} + (y - \bar{y}) \left. \frac{\partial W}{\partial y} \right|_{\bar{x}, \bar{y}},$$

where the partial derivatives are evaluated at $x = \bar{x}$ and $y = \bar{y}$. Thus the covariances in Equations 8.5 and 8.6 can be approximated by

$$\text{cov}(x, W(x, y)) \approx$$

$$E \left\{ (x - \bar{x}) \left(W(\bar{x}, \bar{y}) + (x - \bar{x}) \left. \frac{\partial W}{\partial x} \right|_{\bar{x}, \bar{y}} + (y - \bar{y}) \left. \frac{\partial W}{\partial y} \right|_{\bar{x}, \bar{y}} \right) \right\}$$

$$\text{cov}(x, W(x, y)) \approx G_x \left. \frac{\partial W}{\partial x} \right|_{\bar{x}, \bar{y}} + B_{xy} \left. \frac{\partial W}{\partial y} \right|_{\bar{x}, \bar{y}},$$

where B_{xy} is the genetic covariance between x and y.

the one given for one character shows that

$$\Delta x = \frac{\text{cov}\left(x, W(x, y)\right)}{\bar{W}} \tag{8.5}$$

$$\Delta y = \frac{\text{cov}\left(y, W(x, y)\right)}{\bar{W}}. \tag{8.6}$$

Let B_{xy} indicate the genetic covariance between x and y. G_x still indicates the genetic variance in x. Assuming that selection is weak (see Box 8.3) yields the following expressions for the changes in the mean values of the two characters:

$$\Delta x \approx G_x \left. \frac{\partial \ln W}{\partial x} \right|_{\bar{x}, \bar{y}} + B_{xy} \left. \frac{\partial \ln W}{\partial y} \right|_{\bar{x}, \bar{y}}$$

$$\Delta y \approx B_{xy} \left. \frac{\partial \ln W}{\partial x} \right|_{\bar{x}, \bar{y}} + G_y \left. \frac{\partial \ln W}{\partial y} \right|_{\bar{x}, \bar{y}}.$$

Thus, the change in the mean for each character is the sum of two terms. In each case, one term, beginning G_x or G_y, gives the change due the *direct* effect of variation in that trait on fitness. A second term, beginning B_{xy}, gives the change due to the *correlation* between the two traits and the effect of variation in the other character on fitness. As we will see, this correlated effect of selection can have a profound effect.

8.2 Fisher's runaway process

Fisher first described an evolutionary process that could lead to the evolution of female preferences for exaggerated male traits in his 1930 book. This is still in print, so you can read the paragraph he devotes to the topic. Unless you are a lot smarter than most people, including us, you will have no idea what he is talking about. Perhaps as a consequence, from 1930 to 1980 few gave much credence to Fisher's ideas, nor did many people think much at all about sexual selection.[106] This all changed in 1981 when Russ Lande, then an assistant professor at the University of Chicago, published a simple mathematical model which explained to the vast majority of the human race dumber than R. A. Fisher how the runaway process might work.[107] This paper and a parallel analysis using major gene models by Mark Kirkpatrick[108] set off an explosion of interest in sexual selection that continues today.

8.2.1 Lande's model

Lande modeled a species with two sex-limited quantitative characters.

Male trait. The first is a male trait—think of it as tail length. An individual male's tail length is t, and the mean in the population is $\bar{t}$. Tail length is subject to stabilizing selection, and the optimal tail length is $t = 0$. We can do this without any loss of generality because we are, in effect, specifying the base of measurement. Females also carry the genes for male tail length, but in them these genes are not expressed.

Female preference. The second trait, p, controls female mating preferences. A female with a positive value of p prefers males with larger-than-average values of t. Females with negative values of p prefer males with smaller-than-average values, and females with $p = 0$ mate at random. The larger the absolute value of p, the stronger is the preference. The average preference value in the population is $\bar{p}$.

Using the normal fitness function as before, we obtain the fitness of a male with trait values t and p as

$$W_m(t,p) = e^{a(t-\bar{t})\bar{p}}e^{-ct^2}.$$

The parameter a measures the strength of the mating preference, and the parameter c measures the strength of stabilizing selection. When $a = 0$, female preferences, if they exist, do not affect male fitness. When $c > 0$, large tails reduce male fitness. The bigger c is, the worse off is a male with any tail length $t > 0$.

Since females don't express the male trait, and males are assumed to provide no parental care or any other benefit to females, all females have the same fitness, $W_f = w_0$, and assuming a 1:1 sex ratio, the overall fitness function is $W(t,p) = \frac{1}{2}W_f + \frac{1}{2}W_m$. Thus the recursions for the two means are

$$\Delta\bar{p} = \frac{1}{2}B_{tp}\left.\frac{\partial \ln W}{\partial t}\right|_{\bar{t},\bar{p}} = \frac{1}{2}B_{tp}(a\bar{p} - 2c\bar{t})$$

$$\Delta\bar{t} = \frac{1}{2}G_t\left.\frac{\partial \ln W}{\partial t}\right|_{\bar{t},\bar{p}} = \frac{1}{2}G_t(a\bar{p} - 2c\bar{t}).$$

The first thing we always do is solve for the possible equilibria of the recursion, which exist where $\Delta\bar{p} = \Delta\bar{t} = 0$. There is a recipe for solving a pair of equations of this type, but it is easy to see that when

$$\bar{p} = \left(\frac{2c}{a}\right)\bar{t} \tag{8.7}$$

both difference equations equal zero. Imagine a graph with $\bar{t}$ on the horizontal axis and $\bar{p}$ on the vertical axis. Then Equation 8.7 is the equation of a line through the origin with slope $\frac{2c}{a}$, and any point on the line is an equilibrium.

The next question is: which parts of the line are stable? The answer, it turns out, is either the whole line is stable and the population is free to drift along it, or the whole line is unstable and the population evolves away from the line at an exponential rate. To see why, let's compute the trajectory of a population:

$$\frac{\Delta \bar{p}}{\Delta \bar{t}} = \frac{B_{tp}(a\bar{p} - 2c\bar{t})}{G_t(a\bar{p} - 2c\bar{t})} = \frac{B_{tp}}{G_t}.$$

Thus, if $\bar{t}$ changes one unit as a result of natural and sexual selection, $\bar{p}$ changes $\frac{B_{tp}}{G_t}$ units as a result of the correlated effect of selection. Now, both $\bar{t}$ and $\bar{p}$ increase above the line of equilibrium, and they both decrease below the line of equilibrium. Thus if the slope of the line of equilibrium is less than $\frac{B_{tp}}{G_t}$, the line is unstable, and if it is greater, the line is stable. If female preferences and the male display trait are positively correlated, then an increase in the male trait also leads to an increase in the female preferences as a consequence of the correlated response to selection. If this response is big enough, then the female trait may increase enough to cause the male trait to increase again in the following generation. There is nothing to stop this positive-feedback process, and the character will "run away" indefinitely. This may be clearer if you take a look at Figure 8.3. If the line is stable, the male and female characters can drift along the line leading to any degree of exaggeration as a stable outcome. Of course this will happen more slowly than in the runaway case.

Throughout this discussion we have assumed that there is a positive covariance between the male trait and the female preference. Why should this be the case? The reason is that that females who prefer bigger males will tend to mate with bigger males and therefore their offspring will have both types of genes, and this will give rise to a positive correlation. To

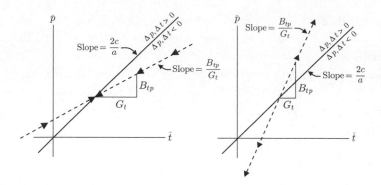

Figure 8.3: A phase plot of the evolutionary trajectories for Lande's model. The solid line gives the possible equilibrium values along which the strengths of natural and sexual selection balance. Above this line both $\bar{p}$ and $\bar{t}$ increase. Below the line both decrease. The dashed lines give the trajectories of populations that begin at different initial points and have slope $\frac{B_{tp}}{G_t}$. (a) If the slope of the trajectories is less than the slope of the line of equilibrium, the system is stable. (b) If the slope of the trajectories is greater, the population evolves stronger and stronger female preferences and more and more exaggerated male characters.

model the evolution of the covariance structure requires much more detailed assumptions about the genetics of the character. Assuming that there are a large number of linked loci with purely additive effects and Gaussian mutation, Barton and Turelli[109] derived expressions for the quantities of interest. Pomiankowski and Iwasa[110] applied these results to the model at hand. They showed that

$$B_{tp} \approx \frac{a}{2} G_t G_p.$$

Major gene models by Mark Kirkpatrick and Pomiankowski also show that nonrandom mating will generate the necessary linkage disequilibrium.

8.3 Costly choice and sensory bias

Lande's model is useful because it exposes the logic of the runaway process in a particularly simple way. However, the model suffers from two related deficiencies:

1. **The runaway process can't really go on forever.** You would have peacocks the size of Jupiter in no time at all. Something must bring the evolution of the male trait to a halt.

2. **Female choice is costless.** This may be approximately true in many cases, but approximately is not good enough in this case. When the line is stable, the male trait stops changing, and the cost of choice, no matter how weak, will govern the evolution of female preferences. When the line is unstable, something must stop the male trait from evolving, and when it does, once again the cost of female choice will predominate.

In this section we will add the costly female choice to Lande's model, and we will see that the costly female choice does undermine the runaway process. In a subsequent section (and in two problems at the end of this chapter), we will see how Fisher can can be "rescued."

Choosing among males requires some kind of search process. Multiple males have to be sampled, and this may take time, compete with feeding, and expose females to predation. Thus, it may be costly for females to choose. To model this situation we assume fitness is now given by

$$W_f(p) = e^{-b(\beta-p)^2}.$$

This equation formalizes the idea that it is easier for females to find some males than others. The parameter b adjusts the cost of female choice. When $b = 0$, the model reduces to Lande's—choice is without cost, and female fitness becomes a constant. The parameter β adjusts the optimal female choice behavior. Remember that $p = 0$ means that a female chooses at random. So, if $\beta = 0$, choosing at random is least expensive. If β is some other value, choosing nonrandomly is best for females.

Why ever let β take any value other than zero? As Mike Ryan argued,[111] it often may be cheaper for females to *have* preferences. One reason is that larger, louder, or brighter males may be easier to find than other kinds of males. Sensory or learning systems may be designed so that some male signals are easier to detect or to process. If this seems implausible, think Las Vegas. Big, bright, colorful signs attract customers because they are attention grabbing. Our sensory systems did not evolve in order to make such signs attractive; they are attractive because our sensory system has responded to many selection pressures, and it has a complex design that serves many functions. A side effect of this design is that big, bright neon signs are more salient than small, tasteful, wooden ones. The same goes for female frogs, except that it is deep croaks, not orange neon, that yank their chains. A second reason is that the act of mating itself may be costly to females, and by having a psychology that makes them receptive only to rare males, females reduce the cost of mating.

Using our general rule, and recalling that $W = \frac{1}{2}W_f + \frac{1}{2}W_m$, we find the resulting recursions

$$\Delta \bar{p} = \frac{1}{2}B_{tp}(a\bar{p} - 2c\bar{t}) + G_p b(\beta - \bar{p})$$

$$\Delta \bar{t} = \frac{1}{2}G_t(a\bar{p} - 2c\bar{t}) + B_{tp}b(\beta - \bar{p}).$$

Once again we can solve for the equilibrium value by setting the change in the means to zero. This yields

$$\frac{1}{2}B_{tp}(a\bar{p} - 2c\bar{t}) - G_p b(\beta - \bar{p}) = 0$$

$$\frac{1}{2}G_t(a\bar{p} - 2c\bar{t}) - B_{tp}b(\beta - \bar{p}) = 0.$$

This is a system of two linear equations in two unknowns, and like almost all such systems, it has a single solution. To find the solution, we could fool around with substitutions and eliminations and all that, but this is one of the cases when just looking at the equations and thinking is the easiest. If you do this, you will eventually see that the solution is $\bar{p} = \beta, \bar{t} = \frac{a}{2c}\beta$. This says that selection adjusts female preferences so that the costs of choice are minimized, and that the male character evolves until the costs of sexual and natural selection are balanced. But here's the interesting part: notice that the solution doesn't depend on b, the parameter that determines the cost of female choice. As long as $b > 0$, the cost of choosing will determine the evolutionary outcome for both males and females, even if it is much, much smaller than the other parameters that determine the effects of choice and selection on males, a and c.

To understand what is going on here, it will be useful to compute the trajectories of $\bar{p}$ and $\bar{t}$ in the same way as before:

$$\frac{\Delta \bar{p}}{\Delta \bar{t}} = \frac{B_{tp}(a\bar{p} - 2c\bar{t}) - G_p b(\beta - \bar{p})}{G_t(a\bar{p} - 2c\bar{t}) - B_{tp} b(\beta - \bar{p})}.$$

If we assume that $b << a, c$, then we can decompose the trajectories of the means into two components (Figure 8.4). When the population is any distance from the equilibrium line, then the population evolves just as it did when there was no cost to female choice. If the covariance is great enough, the system is still unstable. However, when the slope of the line of equilibrium is greater than $\frac{B_{tp}}{G_t}$, the population evolves rapidly until it is close to the line. When it gets there, however, the effects of sexual selection and natural selection on the male character almost balance, and now the costs of female choice come into play. Because choice has small but positive costs, selection gradually adjusts female preferences until $\bar{p} = \beta$, the value that minimizes the costs of choosing for females. Pomiankowski and Iwasa[112] showed that the rate of change of the mean preference along this line is approximately $\Delta \bar{p} \approx G_p b(\beta - p)$. (The methods they used are somewhat beyond the scope of this book.)

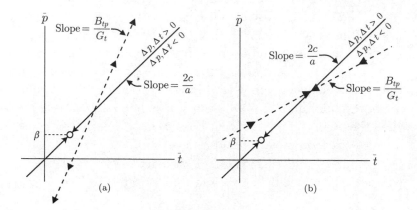

Figure 8.4: A phase plot of the evolutionary trajectories for Lande's ·model with costly female choice added. It is assumed that the costs of choice for females are much smaller than the effects of choice and natural selection on males. The solid line gives the possible equilibrium values along which the effects of natural and sexual selection balance. Above this both $\bar{p}$ and $\bar{t}$ increase. Below the line both decrease. The dashed lines give the trajectories of populations that begin at different initial points and have slope $\frac{B_{tp}}{G_t}$. (a) If the slope of the trajectories is greater than the slope of the line of equilibrium, the system is unstable, and costs of female choice have no effect. (b) If the slope of the trajectories is less than the slope of the line of equilibrium, populations converge to the solid line, and then the costs of female choice shape female preferences so that the costs to females are minimized at $\bar{p} = \beta$.

Okay, but what *is* going on here? Away from the line both traits evolve under the influence of the runaway process just as before. One way of thinking about this is the following: Away from the line, the average male in the population is not the most fit—above the line, the average male character is too small, given the average female preference. This means that males with bigger character values (bigger tails, louder croaks, or whatever) have higher fitness, and the mean value of the male character increases. And since the costs of female choice are small, it also means that females who prefer such males will produce attractive sons and have higher fitness, so female preferences will increase too. Thus, away from the line, selection favors preferences for fathers who will sire "sexy sons." This is an old idea put forward by two bird biologists, Weatherhead and Robertson, in a nonmathematical paper.[113] Below the line, males are too big, and for the same reasons the male character and female preferences will both decrease. Now selection favors females who mate with males who will sire "surviving sons." On the line, sexual selection and natural selection are exactly balanced, so the average male is also the optimal male. This means that there is no advantage to females for picking larger or smaller males. The only thing that matters now is what it costs females to do the choosing. Thus, very weak selective forces acting on female preferences can predominate.

8.4 Good genes and sexy sons

In the last section we saw that minuscule costs determine the evolution of female preferences. At equilibrium it doesn't matter who females mate with—the average male is the best male, and so a preference for larger or smaller males gets you nothing. You might think that this situation is weird, but in fact it characterizes almost every model we have analyzed in this book. In most models there is no heritable variation in fitness at equilibrium. The reason is simple: if there were heritable variation in fitness, it wouldn't be an equilibrium; some types would increase and others decrease. In fact, R. A. Fisher called this principle, the "Fundamental Theorem of Natural Selection." Some population geneticists, especially those who emphasize the nonadaptive complications introduced by multiple genetic loci, are fond of joking that Fisher's Fundamental Theorem is neither fundamental nor a theorem.

Mainly we have ignored such complications, but in this case they are right. Heritable genetic variation can persist for several different reasons:

1. Mutation can maintain heritable variation in fitness. The effect is very small at each genetic locus, but measurements in *Drosophila* suggest that mutation can maintain substantial heritable variation in fitness when averaged over all the loci in the genome.

2. Spatial variation in habitats can maintain heritable variation in fitness because it brings in locally deleterious alleles from elsewhere.

3. Temporal variation in selection can prevent the population from every reaching equilibrium.

If one of these processes maintains heritable variation in fitness, then females can benefit from choosing among males so that their sons or their daughters have higher fitness.

Much of the literature in this area describes such "good-genes" models as alternatives to Fisherian models. In good-genes models, the argument goes, females choose higher-quality males because this will enhance the fitness of their daughters; in Fisherian models females choose males who will produce sexy sons. In a recent paper, Hanna Kokko and her coauthors[114] argued convincingly (to us at least) that this is a false dichotomy. In good-genes models, there is heritable variation that affects survivorship and other components of male and female fitness and no heritable variation in male attractiveness. In Fisherian models, it is the reverse. Good-genes models and Fisherian models are ends of a continuum, and there are many models that mix features of each.

Kokko and her coauthors illustrate their argument with a model of sexual selection that is based in evolutionary game theory. This model builds in a bit more ecological realism than most models, so it is also a bit more complicated. None of the math is difficult, but you have to keep a lot of stuff in mind. If you get lost, just go back and make sure you understand each step.

Unlike most of the models we have seen so far, births and deaths occur in continuous time. This means that individuals are born, live some period of time, and then die. Lifespan may vary among individuals. The sex ratio is fixed at 1:1, so there are always equal numbers of males and females. Both males and females carry a gene that determines their "type." It has two alternative alleles imaginatively labeled $i = 1, 2$. Inheritance is haploid. The gene is only expressed in males, and does not produce observable differences in phenotype. However, males also have an observable display trait that affects mortality differently for type 1 and type 2 males. The mortality of a type i male with display trait D is $\mu_i(D)$, where $\frac{\partial \mu_i(D)}{\partial D} > 0$. This just means that mortality increases as display value increases. We will assume that selection adjusts development so that type 1 and type 2 males express the fitness-maximizing values of D that lead to mortality rates μ_1 and μ_2, respectively. All females have the same death rate μ_f. Mutation maintains heritable variation in fitness. Allele 1 mutates to allele 2 at a rate m_1, and allele 2 mutates to allele 1 at a rate m_2.

Both males and females carry a second gene that affects female mate choice. It is only expressed in females. The strength of female preference is measured by P. Females who carry a preference allele for $P = 0$ mate at random. Females with larger values of P expend effort to mate with males with a larger value of D. In particular, the probability that a female with preference P is impregnated by a male with display value D is $M(P, D)$, where

$$\frac{\partial}{\partial P} \left(\frac{\partial M}{\partial D} \right) > 0.$$

This means that females with larger values of P are more likely to mate with males with larger values of D. Because choice takes time and effort, females with larger values of P have lower fecundity. To represent this assumption, assume that female fecundity, $F(P)$, is such that $\frac{\partial F}{\partial P} < 0$—as preference value increases, fecundity decreases.

We want to use this model to answer two questions:

1. What are the ESS female preferences? What values of P, when common, can resist invasion by individuals carrying genes for all other values, $P' \neq P$. Let $P^\star$ be the ESS value of P.

2. What are the ESS male display traits? If $P^\star$ is common, what are the corresponding frequencies of the male trait D?

To answer these questions, we have to find the value of P that best balances the direct costs of choice in terms of lost fecundity against the indirect benefits that accrue from mating with higher-quality males. Now here's the tricky part. The indirect benefits depend on all descendant generations of both sexes. A female who mates with an attractive male gets indirect benefits because her sons will be more attractive. However, she also gets indirect benefits because her daughters carry the gene that will make their sons more attractive, and so on down the generations. You will remember that we ran into the same problem in Chapter 7 when we dealt with genes that affect sex ratio. We solved it using the reproductive value approach, and we will use the same approach here.

The first thing we have to do is figure out the average lifespan of each type. Assume that virtually all the females in the population have some preference value P and that the frequency of type 1 alleles among newborns is x. Suppose that $2N_0$ individuals are born at time 0, half males and half females. Remember all females have a constant death rate μ_f. Thus, the number of females who remain alive at time t is $N_0 \times \exp(-\mu_f t)$, and the population size of females is proportional to the average lifespan of females, $1/\mu_f$ (Box 8.4). Since female mortality doesn't depend on the male trait, females of all ages are type 1 with probability x. Male mortality does depend on this allele. This means that the population size of type 1 males is proportional to $\frac{x}{\mu_1}$ and the population size of type 2 males is proportional to $\frac{1-x}{\mu_2}$. Notice that, since males and females are born at a constant rate, the average lifespans of females and males of each type are also $\frac{1}{\mu_f}$, $\frac{1}{\mu_1}$, and $\frac{1}{\mu_2}$, respectively. These values will be useful to us because they tell us how long each type lives to reproduce, as well as the relative proportions of each type in the population.

Now we are ready to calculate the indirect benefit of different sorts of mating. The first step is to calculate the type of offspring of males and females of different types. The males are simpler, so let's begin with them. In this model mutation maintains heritable variation in fitness. We assume that mutations from type 1 to type 2 occur with probability m_1 and mutations from type 2 to type 1 occur with probability m_2. Then, from the rules of haploid inheritance, the probability that the offspring of a type 1 male is also type 1, K_1, is

$$K_1 = \underbrace{x(1 - m_1)}_{\text{mates with type 1 female}} + \underbrace{(1 - x)\frac{1}{2}(1 - m_1 + m_2)}_{\text{mates with type 2 female}}.$$

The probability that a type 1 male mates with a type 1 female is x. Such matings produce type 1 offspring with probability $1 - m_1$, that is, as long as there are no mutations. A type 1 male mates with a type 2 female with probability $1 - x$. These matings produce type 1 offspring if the offspring inherits a type 1 allele from dad and there is no mutation— which happens with probability $\frac{1}{2}(1 - m_1)$—or if the offspring inherits a type 2 allele from mom and there is a mutation.

Box 8.4 Mortality and lifespan

The form of mortality assumed in this model is common and useful in many kinds of mathematics, so we explain it in a little more detail here. If the instantaneous mortality rate (sometimes called the "force of mortality") μ is constant with respect to age, then the probability of surviving to age t is given by

$$e^{-\mu t}.$$

This is exponential decay, just like radioactive decay in physics. If the individual survives to age t, the chances of surviving another t years will be the same. We can use this fact to compute the average lifespan of an individual. Since individuals can die at any moment, not just at discrete ages, formally we compute the expected lifespan by taking the integral of the function above:

$$\int_0^\infty e^{-\mu t}dt = \frac{1}{-\mu}\left(e^{-\mu t}\big|_{t=\infty} - e^{-\mu t}\big|_{t=0}\right) = \frac{1}{\mu}.$$

The figure below plots survival to age t for two values of μ, $\mu_1 = 0.25$ and $\mu_2 = 0.10$. The average lifespan in each case is shown by a vertical line.

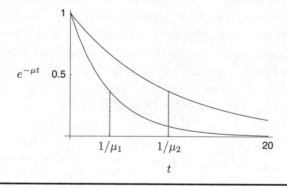

This happens with probability $\frac{1}{2}m_2$. We don't have to worry about genetic correlations between the male character and the female preference because all of the females have preference value P. The same kind of calculation shows that the probability that the offspring of a type 2 male is type 1, K_2, is

$$K_2 = x\frac{1}{2}(1 - m_1) + (1 - x)m_2.$$

Now for the females. Let k_1 be the portion of the offspring of a type 1 females who are type 1, and k_2 the portion of the offspring of a type 2 females who are type 1. The probability that a female mates with a type 1 male, $y(P)$, depends on P, the strength of female preferences, and the frequency of the two types among males. In fact,

$$y(P) = \frac{\frac{x\mu_f}{\mu_1}M(P, D_1)}{\frac{x\mu_f}{\mu_1}M(P, D_1) + \frac{(1-x)\mu_f}{\mu_2}M(P, D_2)} \equiv \frac{x\mu_f}{\mu_1}M_1(P).$$

The term $\frac{x}{\mu_1}$ is proportional to the availability of type 1 males, and $\frac{1}{\mu_f}$ is proportional to the number of females. Thus $\frac{x\mu_f}{\mu_1}$ is the number of type 1 males per female. $M(P, D_1)$ is the probability of mating with such a male when encountered. Thus the numerator gives the number of matings by type 1 males. This is normalized by the sum of both matings to obtain the probability that a given mating is with a type 1 male. We also define $M_1(P)$ as the reproductive success of a single type 1 male, a function that will come in handy in just a bit. The reproductive success of a type 2 male, $M_2(P)$, is defined in the same way. Then we can calculate k_1 and k_2 in the same way that we did with males

$$k_1(P) = \underbrace{y(P)(1 - m_1)}_{\text{mates with type 1 male}} + \underbrace{\left(1 - y(P)\right)\frac{1}{2}(1 - m_1 + m_2)}_{\text{mates with type 2 male}}$$

and

$$k_2(P) = y(P)\frac{1}{2}(1 - m_1) + \left(1 - y(P)\right)m_2.$$

Now we use these values to calculate the equilibrium frequency of type 1 among newborns, $\hat{x}$. Remember that $k_1(P)$

is the fraction of type 1 offspring produced by type 1 females, $k_2(P)$ is the number of type 1 offspring produced by type 2 females, and x is the frequency of type 1 females. Thus,

$$\hat{x} = \hat{x}k_1(P) + (1 - \hat{x})k_2(P). \tag{8.8}$$

Finally, let the reproductive value of a type i male be V_i and that of a type i female be v_i. At equilibrium these satisfy the equations

$$V_1 = \frac{M_1(P)F(P)}{\mu_1}\Big(K_1(V_1 + v_1) + (1 - K_1)(V_2 + v_2)\Big) \tag{8.9}$$

$$V_2 = \frac{M_2(P)F(P)}{\mu_2}\Big(K_2(V_1 + v_1) + (1 - K_2)(V_2 + v_2)\Big) \tag{8.10}$$

$$v_1 = \frac{F(P)}{\mu_f}\Big(k_1(P)(V_1 + v_1) + \big(1 - k_1(P)\big)(V_2 + v_2)\Big) \tag{8.11}$$

$$v_2 = \frac{F(P)}{\mu_f}\Big(k_2(P)(V_1 + v_1) + \big(1 - k_2(P)\big)(V_2 + v_2)\Big). \tag{8.12}$$

Since these give the *relative* value of sons and daughters of the two types, we can set $v_1 = 1$, and then solve the rest. These equations say that the value of an offspring is equal to the number of offspring that they produce weighted by the value of those offspring. So, for example, the value of a type 1 son, V_1, is equal to the rate at which type 1 males produce offspring, $M_1(P)F(P)$, times the average lifespan of type 1 males, $\frac{1}{\mu_1}$, times the number of type 1 and type 2 offspring multiplied by their value. This is where the long run effects of choice on children, grandchildren, and so on are taken into account.

The ESS display values are calculated by finding the value of D_i that maximizes $\frac{M(P,D_i)}{\mu_i(D_i)}$, the rate at which type i males produce offspring times their lifespan. To find the ESS value of P, calculate the fitness of a rare female who has a different strength of preference, P', using these values. This function,

$W(P', P)$, is

$$W(P', P) = \frac{F(P')}{\mu_f} \Big\{ \big[\hat{x} k_1(P') + (1 - \hat{x}) k_2(P') \big] (V_1 + v_1)$$
$$+ \big[\hat{x}(1 - k_1(P')) + (1 - \hat{x})(1 - k_2(P')) \big] (V_2 + v_2) \Big\}.$$
$$(8.13)$$

The fitness of females with preference P' is given by the rate at which such females produce offspring, $F(P')$, times their lifespan, $1/\mu_f$, times the number of each type of offspring that they produce. The ESS value of P is the value such that $W(P, P) > W(P', P)$.

So far we have a general model. To actually get any results we have to specify the functional forms of $\mu_i(D)$, $F(P)$ and $M(P, D)$. In their paper, Kokko et al. used the following function for how female preference and male traits affect probabilities of mating:

$$M(P, D) = e^{(1+D)^P}.$$

This assumption means that mating success is very sensitive to preference values. Relatively modest increases in preference lead to the very large skews in mating success. Fecundity is assumed to be

$$F(P) = 1 - \frac{1}{1 + e^{-(P - 1/E)}}.$$

As is pictured in Figure 8.5(a), the parameter E affects the cost of expressing preferences. As E gets larger, increasing P leads to a bigger decrease in fertility. Finally, the mortality functions $\mu_i(D)$ are

$$\mu_i(D) = \frac{1}{q_i \left(1 - \left(\frac{D}{q_i} \right)^3 \right)}.$$

Figure 8.5(b) shows that increasing display values, D, increases mortality for all males, but this is modulated by genotype. The values q_i for each type i express how costly display

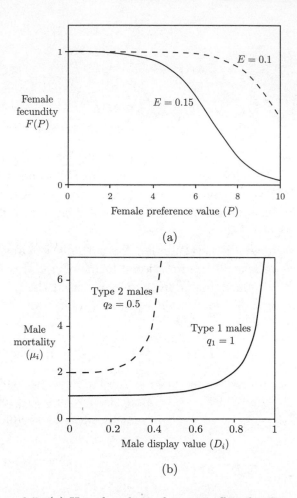

(a)

(b)

Figure 8.5: (a) How female preferences affect fertility. Female fertility is plotted as a function of P, the female preference value. Females with low values of P are not very selective and do not experience a large reduction in fertility. Higher values of P can lead to substantial reductions. Larger values of the parameter E mean that fertility reduction occurs at smaller values of P. (b) How male genotype and display value affects male mortality. Male mortality rate, μ, is plotted as a function of male display value, D, for two parameter values that are used to represent the effect of male genotype in the results plotted in Figure 8.6. Increasing display values increases mortality for both genotypes. However, type 1 males suffer lower mortality than type 2 males with the same display value.

is for a male of that type. High q means that only very high D has much impact on mortality. Smaller q means that even modest D can have a large impact on mortality. Some genotypes, like the type 1 males plotted here, suffer lower mortality than males with other genotypes (like type 2 males) with the same display value. In this example, type 1 males are more able to "afford" a costly display than are type 2 males.

With these functional forms, the model is too complicated to solve analytically. However its can be solved numerically, by assuming specific numerical values of each parameter and then computing the value of P^* and the displays of each type of male. A sketch of the numerical methods used can be found in Appendix D.

The results are shown in Figure 8.6. As long as choice is not too costly, females are quite choosy at equilibrium, and this generates substantial differences in the male display trait for different male genotypes. For example, at low values of E, the ESS value of P is about 3, and the ESS values of D_1 and D_2 are 0.9 and 0.4. Plugging these numbers into $M(P, D)$ reveals that type 1 males have about 60 times higher reproductive success than do type 2 males at equilibrium. However, plugging the ESS value of P into $F(P)$ shows that this level of choice reduces female fertility about 0.1 percent. As female choice gets more costly, selection favors smaller values of P at equilibrium, and the effect of female choice on male fitness is reduced to about a factor of four. The reduction in female fertility due to these preferences is about 1 percent, 10 times larger but still not very large. Having no preference, $P = 0$, is always an ESS, but the invasion barrier necessary for larger values of P to be favored is very small unless choice is quite costly.

This model is interesting because it clearly shows that both Fisherian and good-genes effects can result from *exactly* the same evolutionary processes. You can see this by looking at Figure 8.7, which plots the lifespan of type 1 and type 2 males at evolutionary equilibrium as a function of E, the parameter that determines the costliness of female choice.

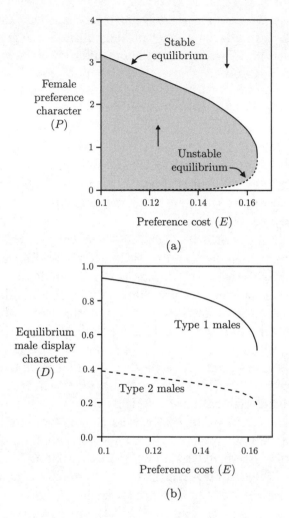

(a)

(b)

Figure 8.6: ESS values of the female preference and male display character. Male mortality is as shown in Figure 8.5(b): $\mu_f = 1$, $m_1 = 0.05$, and $m_2 = 0.01$. (a) The stronger female preferences are favored in the shaded region, and weaker preferences are favored in the unshaded region. The dark line on the top of the shaded region gives the ESS value of P as a function of E. (b) The ESS values of the male display trait for the two male genotypes.

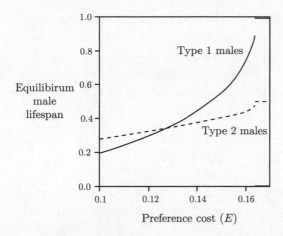

Figure 8.7: The lifespan of type 1 and type 2 males when female preference and male display characters are at their ESS values as a function of E, the cost of female choice. When E is small, type 1 males with large values of D have lower survivorship than type 2 males. Females nonetheless prefer such males because they have greater reproductive success. When E is larger, type 1 males have higher survivorship.

When choice is less costly, the type 1 males with bigger, more costly displays have a shorter lifespan. Here, the model looks Fisherian. Females prefer males with showy displays even though they have lower survivorship because showy displays lead to much greater mating success. When choice is costly, however, it looks like a good-genes model. Selection favors females who prefer type 1 males because they live longer and have somewhat greater mating success. When there is heritable variation in fitness, females can get indirect benefits by choosing males who have "good genes," meaning genes that affect components of fitness like survivorship. However, females can also get indirect benefits by choosing males who have genes that are good because they make male carriers attractive to females.

This model shows that there is no reason to separate these two effects conceptually, and that pure Fisherian models and pure good-genes models are just ends of a continuum. What is happening in any particular population might be a mix of Fisherian and good-genes effects or be closer to one or the other extreme. According to this model at least, it is the costliness of female choice that largely governs how close to either extreme the system lies.

Guide to the Literature

Sexual selection. Andersson 1994 is the most recent book-length overview of the history and content of studies of sexual selection. Kokko et al. 2003 provide a pithy theoretically sophisticated review, and Mead and Arnold 2004 provide an accessible, up-to-date summary of quantitative genetic models of sexual selection. Readers may also find Shuster and Wade 2003 useful. **Quantitative genetics.** The most accessible mathematical introduction we know is in Rice 2004, which discusses the important problem of how genetic variation is maintained. A review by Barton and Keightley 2002 goes into some depth about the problem and provides a thorough bibliography.

Problems

1. Correlated response to selection and evolutionary equilibria. Suppose that individuals are characterized by two morphological characters x and y. Suppose that the fitness of an individual with character values x and y is $W(x, y)$. Show that at equilibrium

$$\frac{\partial \ln W}{\partial x} = \frac{\partial \ln W}{\partial y} = 0,$$

unless the correlation between x and y is exactly 1. How do you explain the fact that the variances and covariances of the trait affect the dynamics of evolution but do not affect the equilibrium?

2. Runaway with biased mutation. Quantitative genetic models all assume that mutation affects the distribution of the genetic values in the population—otherwise selection would rapidly eat up all the variation in the trait being modeled. However, usually it is assumed that the mutation process is unbiased, meaning that it increases the variance but doesn't effect the mean. However, there is every reason to expect that mutation is biased because the amount of bias depends on the scale on which we measure the trait. Suppose that mutation has mean zero and a positive variance when we measure the trait on a linear scale. If we then measure the same trait on a log scale, the same variance will generate a positive mean. Normally we can ignore this effect because we believe that it is small compared to other processes at work. However, in these sexual selection models, once the population reaches the line of equilibrium, the only forces acting on the male trait are very weak ones, and so mutation might have a substantial effect.

In this problem you will show that if biased mutation tends to *decrease* the value of the male trait, weak costs no longer predominate. So, why should it decrease the value of the male trait? The answer is that the male trait is a complex, elaborate thing, and random changes will tend to degrade complex elaborate things. In the same way that mutations are usually deleterious to the complex machinery that organisms require to stay alive, it is also likely to be deleterious to developmental machinery necessary to assemble complex characters like peacock's tails. Measurements of the effect of mutation on fitness show that the effect can be substantial.

(a) Start with the runaway model with costly choice, but add the assumption that mutation decreases the mean value of the male trait an amount μ. To keep things simple, assume that $\beta = 0$ so that unbiased choice is cheapest. Then show that the equations for the change in the mean value of the female preference and the male trait over one generation are, respectively

$$\Delta \bar{p} = \frac{1}{2} B_{tp}(a\bar{p} - 2c\bar{t}) - G_p b\bar{p}$$

$$\Delta \bar{t} = \frac{1}{2} G_t(a\bar{p} - 2c\bar{t}) - B_{tp} b\bar{p} - \mu.$$

(b) Show that the equilibrium values of the male display, $\hat{t}$, and the female preference, $\hat{p}$, are

$$\hat{t} = \frac{\mu \left(a B_{tp} - 2b G_p \right)}{2cb \left(G_t G_p - B_{tp}^2 \right)}$$

$$\hat{p} = \frac{\mu B_{tp}}{b \left(G_t G_p - B_{tp}^2 \right)}.$$

(c) Assuming that the costs of female choice are small, meaning that b is small, prove that the equilibrium values lie close to the line of equilibrium in the problem without mutation. (d) Assume that the means are no longer 0, but instead are proportional to μ. Does that mean that they have to be close to zero and that the effect of mutation on the mean character is also small?

3. When selection limits the runaway process. Even when costs and biased mutation are introduced, Lande's model is still unstable if $\frac{B_{tp}}{G_t} > \frac{2c}{a}$, and this can't be right. NO peacocks ever get as big as a Buick, let alone Jupiter. Adding good genes won't help—so-called "good-genes models" can still be unstable. We know that eventually natural selection can overcome any amount of female choice and halt the runaway, but this idea can't be represented with the assumed fitness function. One way to fix the model is to modify the male viability selection in the basic model so that the fitness of males is

$$W_m(t, p) = e^{a(t-\bar{t})\bar{p}} e^{-ct^4}.$$

Natural selection still has an optimum at $t = 0$, but because t is raised to the fourth power, it is weak for values near the optimum but becomes much stronger for large deviations. (a)

Assuming that the fitness of females is $W_f(t,p) = e^{-bp^2}$, show that the changes in the means are given by

$$\Delta\bar{p} = \frac{1}{2}B_{tp}(a\bar{p} - 4c\bar{t}^3) - G_p b\bar{p}$$

$$\Delta\bar{t} = \frac{1}{2}G_t(a\bar{p} - 4c\bar{t}^3) - B_{tp}b\bar{p}.$$

(b) Assume that $b = 0$. Show that there is a line of equilibrium given by the equation $\bar{p} = \frac{4c}{a}\bar{t}^3$. (c) Show that the trajectories of $\bar{p}$ and $\bar{t}$ in phase space are straight lines with slope $\frac{B_{tp}}{G_t}$. (d) Plot the line of equilibrium and the trajectories in phase space and determine which segments of the line are stable and which are unstable. (e) If b is a small, positive number, then once the population is on a stable segment of the line, it will move toward $\bar{p} = 0$, $\bar{t} = 0$. Assuming that this movement along the line is slow compared to the rates of change when the population is not near the line of equilibrium, sketch the trajectory of $\bar{t}$ as a function of time.

Notes

[101] Julian Huxley 1938a, 1938b. Julian Huxley, a noted biologist, was T. H. Huxley's grandson. His brother Aldous was author of *Brave New World*.

[102] Andersson's 1994 book *Sexual Selection* takes this position, as well as advocates a taxonomy based partly on behavior mechanisms.

[103] Iwasa et al. 1991. Alternative approaches have been developed by Lande and by Barton and Turelli, among others.

[104] Barton and Turelli 1991.

[105] See Bulmer 1985 or Barton and Turelli 1989.

[106]There are some notable exceptions. See Andersson's 1994 book for a historical overview of the topic.

[107]Lande 1981.

[108]Kirkpatrick 1982.

[109]Barton and Turelli 1991.

[110]Pomiankowski and Iwasa 1993.

[111]Ryan 1998.

[112]Pomiankowski and Iwasa 1993.

[113]Weatherhead and Robinson 1979.

[114]Kokko et al. 2002.

Box 8.5 Symbols used in Chapter 8

N Number of individuals in the population

$n(x)$ Number of individuals with trait value x

$\bar{x}$ Average value of trait x

G_x Variance in trait x

θ Optimal trait value with respect to individual fitness

S Width of selection function; how quickly fitness drops as trait value moves away from θ

B_{xy} Covariance between traits x and y

t Value of a male's display trait

p Value of a female's preference trait

a Strength of female mating preference

c Strength of individual selection on male trait value

b Cost of female choice

β Value of optimal female preference with respect to costs of choice

Kokko et al. model:

μ_i Mortality rate of a type i male

μ_f Mortality rate of a female

D Value of male display

m_i Rate at which allele i, governing male type, mutates to the other allele

P Strength of female preference

x Frequency of type 1 alleles among newborns

K_i, k_i Probabilities that an offspring of a type i male (K) or a female (k) is also type i

V_i, v_i Reproductive values of type i male and female, respectively

E Costliness of female preferences

q_i Ability of a male to produce displays without suffering through mortality

Appendix A

Facts about Derivatives

The derivative of a function $y = f(x)$ can be written as $f'(x)$, $y'(x)$, y', or $\frac{dy}{dx}$. With recursions, it is safer to use the dy/dx notation to avoid confusion with the recursion of state variables, which is often denoted by primes. Below are some common differentiation rules with which you should familiarize yourself.

The derivative of a constant is zero

$f(x) = c$, where c is a constant: $f'(x) = 0$.
$f(x) = cg(x)$, where g is a function of x: $f'(x) = cg'(x)$.

Sum rule

The derivative of a sum of two functions is the sum of the derivatives of each:

$$f(x) = g(x) + h(x): \qquad f'(x) = g'(x) + h'(x).$$

Product rule

The derivative of the product of two functions is the sum of the derivative of the first times the second and the derivative of the second times the first:

$$f(x) = g(x) \times h(x): \qquad f'(x) = g'(x)h(x) + h'(x)g(x).$$

Quotient rule

The derivative of the quotient of one function divided by a second is the difference between the derivative of the first times the second and the derivative of the second times the first, divided by the square of the second:

$$f(x) = \frac{g(x)}{h(x)}: \qquad f'(x) = \frac{g'(x)h(x) - h'(x)g(x)}{h(x)^2}.$$

Power rule

The derivative of a variable to the power n is n times the variable to the power $n - 1$:

$$f(x) = x^n: \qquad f'(x) = nx^{n-1}.$$

Taylor expansion

Any function $f(x)$ can be decomposed such that its value at $x + \delta$ is given by

$$f(x + \delta) = f(x) + \delta f'(x) + \frac{1}{2}\delta^2 f''(x) + \ldots.$$

Thus when δ^2 is small enough to ignore, we can approximate $f(x + \delta)$ with a linear approximation of the dynamics: $f(x + \delta) \approx f(x) + \delta f'(x)$.

Appendix B

Facts about Random Variables

A random variable, as we use it in this book, is a value that has yet to be determined by some process. It could take on a number of values, but we want to say something about the statistics of the variable before any specific experiment or process determines the realized value.

Let x be a random variable with mean $\bar{x} = \mathrm{E}(x)$ and y a random variable with mean $\bar{y} = \mathrm{E}(y)$. The expectation of a sum equals the sum of the expectations:

$$\mathrm{E}(x + y) = \mathrm{E}(x) + \mathrm{E}(y).$$

The expectation of ax, where a is a constant, equals a times the expectation of x:

$$\mathrm{E}(ax) = a\mathrm{E}(x).$$

The variance of x is the expectation of the squared deviations from the mean:

$$\mathrm{var}(x) = \mathrm{E}\left\{(x - \bar{x})^2\right\}.$$

This can be expressed also as the expectation of the square minus the squared expectation:

$$\mathrm{E}\left\{(x - \bar{x})^2\right\} = \mathrm{E}(x^2) - \mathrm{E}(x)^2.$$

The covariance of x and y is the expectation of the product of the two minus the product of their expectations:

$$\text{cov}(x, y) = \text{E}(xy) - \text{E}(x)\text{E}(y).$$

Like an expectation, a covariance distributes over summation. Let z be a third random variable. Then,

$$\text{cov}(x + z, y) = \text{cov}(x, y) + \text{cov}(z, y).$$

Constants also factor out:

$$\text{cov}(ax, by) = ab \times \text{cov}(x, y).$$

The slope of the regression line of y on x is the covariance of x and y divided by the variance of x (the predictor variable):

$$\beta(y, x) = \frac{\text{cov}(x, y)}{\text{var}(x)}.$$

The correlation of two random variables is their covariance divided by the square root of the product of their variances:

$$r_{x,y} = \frac{\text{cov}(x, y)}{\sqrt{\text{var}(x)\text{var}(y)}}.$$

Appendix C

Calculating Binomial Expectations

Whenever a game model requires individuals to interact with others in groups greater than two, calculating a strategy's fitness often requires calculating a binomial expectation (like the n-person reciprocity model in Chapter 4). The binomial distribution is given by

$$p_{n,x} = \frac{n!}{x!(n-x)!} p^x (1-p)^{n-x},$$

where p is the probability of an event and $p_{n,x}$ is the probability of getting x events (each with probability p) in n trials. The mean of the distribution is

$$np$$

and the variance is

$$np(1-p).$$

To translate this to the language of game theory models, n is the size of the group within which individuals interact, and x is the number of members of the group of size n who are of the strategy with frequency p. Usually, we know the strategy of at least one member of the group, and that is the strategy we are figuring a fitness for. Thus, the relevant sample size is $n - 1$ instead of n, and we are interested in how many of those $n - 1$ are of a particular strategy. That comes straight from the binomial distribution, but we have to multiply each payoff for that number of each kind of strategy in the group by its probability of occurring. This isn't always easy. Let's start with something easy to show the general tactic.

Example: the mean

To see the general strategy for calculating binomial expectations, let's prove that the mean of the binomial distribution is np. Imagine we want to know on average how many of $n - 1$ individuals in a randomly formed group will be of a strategy with frequency p in the population as a whole. We want to calculate

$$\mathrm{E}(x) = \sum_{x=0}^{n-1} \frac{(n-1)!}{x!(n-1-x)!} p^x (1-p)^{n-1-x} \{x\}.$$

That is, we sum up a weighted average of chances of occurrence of each number of individuals of the strategy. We could rewrite the above as

$$\mathrm{E}(x) = \sum_{x=0}^{n-1} \mathrm{Pr}(x)\{x\},$$

where $\mathrm{Pr}(x) = p_{n-1,x}$ is the chance of seeing x individuals of the strategy among $n - 1$ individuals. Here is how to prove the mean in this case is $(n - 1)p$. First, distribute the x at the end *into* the factorial:

$$\mathrm{E}(x) = \sum_{x=0}^{n-1} \frac{(n-1)!x}{x!(n-1-x)!} p^x (1-p)^{n-1-x}.$$

$$\mathrm{E}(x) = \sum_{x=0}^{n-1} \frac{(n-1)!}{(x-1)!(n-1-x)!} p^x (1-p)^{n-1-x}.$$

Now multiply everything by p and $1/p$ and distribute the $1/p$ into the summation:

$$\mathrm{E}(x) = p \sum_{x=0}^{n-1} \frac{(n-1)!}{(x-1)!(n-1-x)!} p^{x-1}(1-p)^{n-1-x}.$$

Similarly, multiply by $(n-1)$ and $1/(n-1)$:

$$\mathrm{E}(x) = (n-1)p \sum_{x=0}^{n-1} \frac{(n-2)!}{(x-1)!(n-1-x)!} p^{x-1}(1-p)^{n-1-x}.$$

We know that the sum of the probabilities is one (1), like so:

$$\sum_{j=0}^{m} \frac{m!}{j!(m-j)!} p^{j}(1-p)^{m-j} = 1.$$

Using this fact, rewrite the expectation as

$$\mathrm{E}(x) =$$
$$(n-1)p \sum_{x=0}^{n-1} \frac{(n-2)!}{(x-1)!(n-2-[x-1])!} p^{x-1}(1-p)^{n-2-[x-1]}.$$

Now let $j = (x-1)$ and let $m = (n-2)$:

$$\mathrm{E}(x) = (n-1)p \sum_{j=-1}^{m} \frac{m!}{j!(m-j)!} p^{j}(1-p)^{m-j}.$$

Since $(-1)!$ is undefined, the sum has no value at $j = -1$, and the foregoing simplifies to

$$\mathrm{E}(x) = (n-1)p \left\{ \sum_{j=0}^{m} \frac{m!}{j!(m-j)!} p^{j}(1-p)^{m-j} \right\}.$$

The sum in braces is equal to one (1), as before, and we now have

$$\mathrm{E}(x) = (n-1)p \left\{ 1 \right\} = (n-1)p.$$

The general strategy is just like that: distribute into the factorials the quantity to take the expectation of, and then multiply the entire sum by quantities of p and n such to make the part inside the sum equal to one. There is one more trick, though, shown in the next section.

Something harder

Let's try

$$E\left(\frac{1}{x+1}\right) = \sum_{x=0}^{n-1} \frac{(n-1)!}{x!(n-1-x)!}p^x(1-p)^{n-1-x}\left\{\frac{1}{x+1}\right\}.$$

This is a very common expectation, as it occurs whenever we have a "lottery" in which the focal individual (the 1 in the denominator) and all other individuals of the same type participate. That is, if there are x individuals and the focal individual who will share out some resource (be it mates or food or territory), then each gets a share $1/(x+1)$. Thus it is good to know the sum of this expectation, and we will derive it here.

As before, move the fraction on the far right into the factorial:

$$E\left(\frac{1}{x+1}\right) = \sum_{x=0}^{n-1} \frac{(n-1)!}{(x+1)!(n-1-x)!}p^x(1-p)^{n-1-x}.$$

Now multiply by p/p and n/n, moving factors into the sum until we can make the factorial whole again:

$$E\left(\frac{1}{x+1}\right) = \frac{1}{p}\sum_{x=0}^{n-1} \frac{(n-1)!}{(x+1)!(n-1-x)!}p^{x+1}(1-p)^{n-1-x}$$

$$= \frac{1}{np}\sum_{x=0}^{n-1} \frac{n!}{(x+1)!(n-[x+1])!}p^{x+1}(1-p)^{n-[x+1]}.$$

Let $j = (x+1)$:

$$E\left(\frac{1}{x+1}\right) = \frac{1}{np}\left\{\sum_{j=1}^{n} \frac{n!}{j!(n-j)!}p^j(1-p)^{n-j}\right\}.$$

Now the part in braces is equal to 1 minus the probability

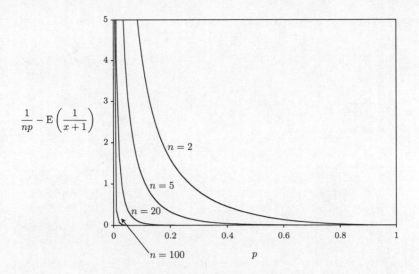

$$\frac{1}{np} - \mathrm{E}\left(\frac{1}{x+1}\right)$$

Figure C.1: The difference between $1/(np)$ as an estimate of $\mathrm{E}(1/(x+1))$ and the actual expectation. As this difference approaches zero, $1/(np)$ is a better estimate of the actual sum. Unless p is very small, this is the case for large groups, but the deviation can be very large for small groups, even when p is large.

of seeing zero successes in n trials, i.e.,

$$\left\{\sum_{j=1}^{n} \frac{n!}{j!(n-j)!}p^j(1-p)^{n-j}\right\} = 1 - \frac{n!}{0!(n-0)!}p^0(1-p)^{n-0}$$

$$= 1 - (1-p)^n.$$

Thus the entire expression is now

$$\mathrm{E}\left(\frac{1}{x+1}\right) = \frac{1}{np}\left\{1 - (1-p)^n\right\} = \frac{1-(1-p)^n}{np}.$$

That's about all the tricks.

Note that when n is large (say about 100 or more) and p is not very small, the above is approximately

$$\frac{1}{np}.$$

This means as groups become larger, $1/(np)$ becomes a better approximation of $E(1/(x + 1))$. For small groups, the exact expression above is necessary (see Figure C.1).

Appendix D

Numerical Solution of the Kokko et al. Model

Very little of theory presented in this book requires any numerical work to get results. The Kokko et al. (2002) model presented in Chapter 8 is an exception. In this appendix we briefly sketch the numerical methods that are required to solve it. The same methods can be used to compute ESS strategies for many models that are too complex to get complete analytical solutions for.

Numerical work requires a completely new mind set. Analytical work is all about manipulating symbols. For example, to solve the Kokko et al. model we have to find the value of the male display trait D that maximizes the function

$$\frac{M(P,D)}{\mu_i D} = q_i \left(1 - \left(\frac{D}{q_i}\right)^3\right) e^{(1-D)^P}.$$

The analytic approach is to set the partial derivative of this function with respect to D equal to zero and manipulate the resulting equation until the symbol D is alone on one side of the equation. The numerical approach is to pick numerical values for the parameters μ_i and P and then try a lot of different values of D looking for the one that yields the biggest value of the function. The art of numerical work is to choose

343

an algorithm that picks the values of D so that we converge rapidly and reliably on the value that maximizes the function. There are lots of different algorithms and which one is best depends on the details of the problem.

In this case, we are trying to find the maximum of a function of a single variable. Moreover, studying the analytic form of the function tells us that we only need to try values of D between 0 and μ_i because larger values lead to negative reproductive success, and (by differentiating it) that the function only has one local maximum in the interval. Maximizing a unimodal function of a single variable is an easy numerical problem, and there are a number of very efficient algorithms. One of these, called *golden section* search, is illustrated in Figure D.1. If the function had more than one variable or had many local maxima, the problem would be a bit trickier, but with a little care could still be solved. An excellent discussion of numerical methods for maximizing functions can be found in *Numerical Recipes*, a wonderful series (different versions for C, C++, and Fortran) of texts on numerical methods that are available free online at www.numerical-recipes.com.

With this introduction in mind, we sketch the steps necessary to produce the results shown in Figure 8.6. The problem is to find the values of P that satisfy the equation

$$H(P) = \left. \frac{\partial W(P', P)}{\partial P'} \right|_{P'=P} = 0,$$

where $W(P', P)$ is defined in Equation 8.13 as a function of the variable E. $D = D^\star$ is the value that maximizes $\frac{M(P,D)}{\mu_i D}$. Once again the essence of the problem is to choose values of P until you find one that satisfies $H(P) = 0$, and once again the art is to find one that does this both reliably and rapidly. This particular problem is a bit tricky because there are sometimes two values that satisfy the equation and sometimes none. There are many good algorithms, and you should consult *Numerical Recipes* for the details.

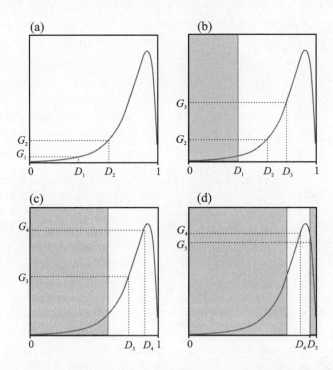

Figure D.1: The first four steps in finding the value of D that maximizes $G(D) = \frac{M(P,D)}{\mu_i D}$. The figure plots this function with $P = 3$ and $\mu_i = 1$ over the interval $[0, 1]$. (a) The first step is to calculate the value of the function at two values of D, $D_1 = 0.38$ and $D_2 = 0.72$. These values are called the "golden ratio" and play a big role in art and architecture. We'll see the reason for choosing these particular values in a moment. As you can see, $G(D_2) > G(D_1)$. Since $G(D)$ is unimodal, this means that the value of D that maximizes $G(D)$ is greater than D_1. (b) The next step is to choose a new value of D, D_3, which lies a fraction 0.72 of the way between D_1 and 1 and compute $G_3 = G(D_3)$. Because of our choice of the golden ratio fractions, D_2 now lies a fraction 0.38 of the way between D_1 and 1. So, since $G_3 > G_2$, we know that the maximizing value of D is greater than D_2. (c) Repeating same procedure tells us that the maximizing value of $D > D_3$. (d) Now $G(D_5) < G(D_4)$, which tells us that the maximum lies in the interval $[D_3, D_5]$. As we repeat this algorithm, the interval in which the maximimum lies decreases by a fraction 0.72 for each evaluation of the function $G(D)$ and thus converges geometrically to the optimum.

Any of these algorithms will need to be supplied with values of $H(P)$. This will require several steps.

(1)   The first step is to compute $H(P) = \left.\frac{\partial W(P',P)}{\partial P'}\right|_{P'=P}$. This is

$$
\begin{aligned}
H(P) = \left.\frac{\partial F(P')}{\partial P'}\right|_{P'=P} & \left\{ \left[\hat{x}k_1(P) + (1 - \hat{x})k_2(P)\right](V_1 + v_1) \right. \\
& + \left[\hat{x}(1 - k_1(P)) + (1 - \hat{x})(1 - k_2(P))\right](V_2 + v_2) \Big\} \\
& + F(P)\Big\{ \hat{x}\left.\frac{\partial k_1(P')}{\partial P'}\right|_{P'=P} (V_1 - V_2 + v_1 - v_2) \\
& + (1 - \hat{x})\left.\frac{\partial k_2(P')}{\partial P'}\right|_{P'=P} (V_1 - V_2 + v_1 - v_2) \Big\}
\end{aligned}
$$

$$(D.1)$$

The reproductive values V_1, V_2, v_1, and v_2 can be found by setting any one of them to 1 and then solving the remaining three simultaneous linear equations 8.12. Similarly the equilibrium frequency of type 1 offspring is found by solving Equation 8.8 for $\hat{x}$. The partial derivatives of k_1 and k_2 are found by substituting in the functional forms for $M(P', D)$, $D(P')$, and $\mu_i(D)$ and differentiating with respect to P' and finally evaluating them at P and $D^\star$. Remember that $D^\star$ is the value of D that maximizes $G(P, D)$ for the given value of P (found using a numerical optimization routine like golden section search).

(2)   Once you have $H(P)$, you next choose numerical values for the parameters. The plotted results assume that $\mu_1 = 1$, $\mu_2 = 0.5$, $\mu_f = 1$, $m_1 = 0.05$, and $m_2 = 0.01$. Then:

1. Set $E = 0.1$.

2. Use a numerical root-finding routine to find the value of P that satisfies $H(P) = 0$ for that value of E. As part of this calculation you will also find $D^\star$ for that value of E.

3. Increment E by the desired accuracy (say 0.001) and repeat until $E = 0.16$.

Whew. With the series of values for P and $D^\star$ that you get from the above procedure, you can construct the plots in Figure 8.6.

Appendix E

Solutions to Problems

Chapter 1

1. Linear dynamics

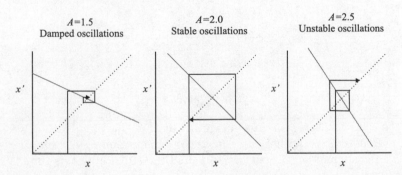

$A=1.5$
Damped oscillations

$A=2.0$
Stable oscillations

$A=2.5$
Unstable oscillations

2. Two-way mutation

To find the equilibrium value, $\hat{p}$, solve

$$\hat{p} = \hat{p} - m_{12}\hat{p} + m_{21}(1 - \hat{p}).$$

Subtracting $\hat{p}$ from both sides and collecting terms gives

$$0 = \hat{p}(-m_{12} - m_{21}) + m_{21}.$$

Now add $\hat{p}(m_{12} + m_{21})$ to both sides:

$$\hat{p}(m_{12} + m_{21}) = m_{21}.$$

Dividing by $(m_{12} + m_{21})$ yields the desired expression. ∎

3. Social learning and replicator dynamics

With the assumptions given in the problem, we can write down the following mating table:

Self	Other	Probability	$\Pr(A)$	$\Pr(B)$
A	A	p^2	1	0
A	B	$p(1-p)$	$1-e$	e
B	A	$(1-p)p$	$1-e$	e
B	B	$(1-p)^2$	0	1

Thus
$$p' = p^2(1) + 2p(1-p)(1-e) + (1-p)^2(0).$$

Simplifying gives

$$
\begin{aligned}
p' &= p^2(1) + 2p(1-p) - e2p(1-p) \\
&= p^2 + 2p - 2p^2 - e2p(1-p) \\
&= p + p - p^2 - e2p(1-p) \\
&= p + p(1-p)(1-2e).
\end{aligned}
$$

Subtracting p from both sides yields the desired form. ∎

4. When fitness is not a constant

(a) With the assumtion that individuals in this population mate at random, the probability that an A gamete unites with another A gamete and has fitness w_{AA} is p, and the probability this gamete unites with an S-bearing gamete and has fitness w_{AS} is $1-p$. Thus the average fitness of an A gamete is

$$W_A = pw_{AA} + (1-p)w_{AS}.$$

By the same reasoning,

$$W_S = pw_{AS} + (1-p)w_{SS}.$$

(b) According to the difference equation given above, the population can be at equilibrium only if $W_A = W_S$. Substituting the expressions derived in part (a) yields

$$\hat{p}w_{AS} + (1-\hat{p})w_{SS} = \hat{p}w_{AA} + (1-\hat{p})w_{AS}.$$

Collecting terms gives

$$\hat{p}(w_{AS} - w_{SS}) + w_{SS} = \hat{p}(w_{AA} - w_{AS}) + w_{AS}$$

or

$$\hat{p}(2w_{AS} - w_{SS} - w_{AA}) = w_{AS} - w_{SS}.$$

Solving for $\hat{p}$ yields the desired expression.

(c)

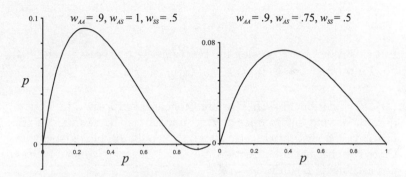

(d) From part (b) there is only one equilibrium. From Chapter 1, the slope of the recursion at $p = 0$ is $\frac{W_A}{W_S} \approx \frac{w_{AS}}{w_{SS}}$. Similarly, the slope of the recursion at $p = 1$ is $\frac{W_S}{W_A} \approx \frac{w_{AS}}{w_{AA}}$. Thus if $w_{AS} > w_{SS}, w_{AA}$ both of these points are unstable. This means that the recursion must cross the 45-degree line once with a slope less than 1. As long as selection is not too strong the slope will be greater than -1. ∎

5. When environments vary with time

(a) Suppose at time $t = 0$, during the wet season, the population has size N_0 and that the frequency of A is p. Then the number of A types at the beginning of the dry season is

$$N_{A1} = N_{A0}$$

and the number of B types is

$$N_{B1} = 0.2 N_{B0}.$$

The numbers at the beginning of the wet season during the next year are

$$N_{A2} = N_{A1} = N_{A0},$$

$$N_{B2} = 0.2N_{B1} = 0.2(2N_{B0}) = 0.4N_{B0}.$$

Thus the frequency of the A type after 1 year is

$$p' = \frac{N_{A2}}{N_{A2} + N_{B2}} = \frac{N_{A0}}{N_{A0} + 0.4N_{B0}}.$$

Dividing by $N_{A0} + N_{B0}$ yields

$$p' = \frac{p}{p + 0.4(1 - p)}.$$

But this is just the recursion for two types with fitnesses 1 and 0.4. So A always increases.

(b) Population growth is a multiplicative process. In a temporally varying environment, what matters is the product of the fitnesses, not their arithmetic average. Population biologists sometimes say that natural selection maximizes *geometric* mean fitness, not arithmetic mean fitness. ∎

6. Horizontal cultural transmission, again

With the assumptions given in the problem, we can write down the following mating table:

Self	Other	Probability	Pr(A)	Pr(B)
A	A	p^2	1	0
A	B	$p(1-p)$	1	0
B	A	$(1-p)p$	$\beta(V(A) - V(B))$	$1 - \beta(V(B) - V(A))$
B	B	$(1-p)^2$	0	1

Thus

$$p' = p^2(1) + p(1 - p)(1) + p(1 - p)\beta(V(A) - V(B)) + (1 - p)^2(0).$$

Simplifying gives

$$p' = p^2(1) + p(1 - p)(1) + p(1 - p)\beta(V(A) - V(B)) + (1 - p)^2(0)$$
$$= p^2 + p - p^2(1) + p(1 - p)\beta(V(A) - V(B)).$$

Subtracting p from both sides yields the desired form. ∎

Chapter 2

1. Conflict as cooperation

(a) $V(R|R) = 0 < V(W|R)$ only if $k < b$. We have $0 = V(W|R) < b - \frac{k}{2} = V(W|W) < b = V(R|W)$ as long as $b, k > 0$.

(b) The fitness of Rest is

$$V(R) = pV(R|R) + (1-p)V(R|W) = (1-p)b.$$

The fitness of Wash is

$$V(W) = pV(W|R) + (1-p)V(W|W) = p(b-k) + (1-p)\left(b - \frac{k}{2}\right).$$

At equilibrium these fitnesses must be equal, so the value of $\hat{p}$ must satisfy

$$(1-\hat{p})b = \hat{p}(b-k) + (1-\hat{p})(b - \frac{k}{2})$$

$$0 = \hat{p}\left(b - k - \left(-\frac{k}{2}\right)\right) - \frac{k}{2}$$

$$\hat{p} = \frac{k}{2b - k}.$$

(c) Introducing Spiteful yields the following payoff matrix:

	Rest	Wash	Spiteful
Rest	0	b	0
Wash	$b-k$	$b - \frac{k}{2}$	$b - \frac{k}{2}$
Spiteful	0	$b - \frac{k}{2}$	$b - \frac{k}{2}$

Spiteful can resist invasion by rare Rest individuals when $V(S|S) > V(R|S)$ or $b - \frac{k}{2} > 0$. Spiteful can never resist invasion by Wash because $V(S|S) = V(W|S)$. The fitnesses of R and W are the same when S is rare, and $p = k/(2b - k)$ because it is an equilibrium when those are the only strategies present. Rare S individuals will be able to invade if $V(S) > V(R) = V(W)$. Thus

$$V(S) = \hat{p}(0) + (1 - \hat{p})(b - \frac{k}{2}) > (1 - \hat{p})b = W(R),$$

as long as $k > 0$. Spiteful is like Retaliator in that it punishes Rest individuals but pays the cost of foregoing the benefits of solo washing. ∎

2. Display costs

(a) Here is the payoff matrix:

	Hawk	Dove
Hawk	$\frac{v-c}{2}$	v
Dove	0	$\frac{v-d}{2}$

Hawk is not an ESS for the same reason as in the usual game. Dove is not an ESS because $v > \frac{v-d}{2}$.

(b) Setting the average payoffs of the two strategies equal yields

$$\hat{p}\frac{v-c}{2} + (1-\hat{p})v = (1-\hat{p})\frac{v-d}{2}$$
$$\hat{p}(v-c) + (1-\hat{p})(v+d) = 0$$
$$(v+d) = \hat{p}(-v+d+v+c)$$
$$\frac{v+d}{c+d} = \hat{p}.$$

(c) No, the only effect is to increase the frequency of Hawks at the internal equilibrium. ∎

3. More display costs

(a) The payoff matrix in this game becomes

	Hawk	Dove	Bourgeois
Hawk	$\frac{v-c}{2}$	v	$\frac{v}{2}+\frac{v-c}{4}$
Dove	0	0	0
Bourgeois	$\frac{v-c}{4}$	$\frac{v}{2}$	$\frac{v}{2}$

When Bourgeois is absent the game is as in Problem 1. Bourgeois is an ESS since $\frac{v}{2} > 0$.

(b) The same argument given in the text shows that Bourgeois have the same fitness at the H-D equilibrium as Hawks and Doves.

(c) No. ∎

4. Correlated versus uncorrelated asymmetries

(a) We already showed that the payoffs to each strategy against itself are

$$v(B|B) = V(A|A) = \frac{v}{2}.$$

The payoff to a Bourgeois against an Assessor is

$$V(B|A) = \frac{1}{2} \left\{ \underbrace{\underbrace{\frac{v}{2}}_{\text{B is bigger}} + \underbrace{\frac{(1-x)v - xc}{2}}_{\text{B is smaller}}}_{\text{B is owner}} \right\}$$

$$+ \frac{1}{2} \left\{ \underbrace{\underbrace{\frac{v}{2}}_{\text{B is bigger}} + \underbrace{\frac{1}{2}(0)}_{\text{B is smaller}}}_{\text{B is intruder}} \right\}.$$

Likewise, the payoff to an Assessor against a Bourgeois is

$$V(A|B) = \frac{1}{2} \left\{ \underbrace{\underbrace{\frac{v}{2}}_{\text{A is owner}} + \underbrace{\frac{xv - (1-x)c}{2}}_{\text{A is intruder}}}_{\text{A is bigger}} \right\}$$

$$+ \frac{1}{2} \left\{ \underbrace{\underbrace{\frac{v}{2}}_{\text{A is owner}} + \underbrace{\frac{1}{2}(0)}_{\text{A is intruder}}}_{\text{A is smaller}} \right\}.$$

Thus Assessor can invade a population of Bourgeois when

$$\frac{1}{2} \left\{ \frac{v}{2} + \frac{xv - (1-x)c}{2} \right\} + \frac{1}{2} \left\{ \frac{1}{2}\frac{v}{2} \right\} > \frac{v}{2}.$$

or, solving for x,

$$x > \frac{v + 2c}{2(v + c)}.$$

A similar calculation shows that Bourgeois can invade Assessor when

$$x < \frac{v}{2(v + c)}.$$

(b) When both Bourgeois and Assessor are ESSs, there is an unstable mixed equilibrium. To find this equilibrium, let p be the frequency of Assessor in a population comprised of just these two strategies. The fitness of Assessor is then

$$W(A) = w_0 + p\frac{v}{2} + (1-p)V(A|B).$$

The fitness of Bourgeois is similarly

$$W(A) = w_0 + (1-p)\frac{v}{2} + pV(B|A).$$

To find the equilibrium, set $W(A) = W(B)$ and solve for $\hat{p}$:

$$\hat{p}\frac{v}{2} + (1-\hat{p})V(A|B) = (1-\hat{p})\frac{v}{2} + \hat{p}V(B|A)$$

$$\hat{p}\left(\frac{v}{2} - V(B|A)\right) = (1-\hat{p})\left(\frac{v}{2} - V(A|B)\right)$$

$$\hat{p} = \frac{\frac{v}{2} - V(A|B)}{V(A|B) - V(B|A)}.$$

Substituting in the expressions for $V(B|A)$ and $V(A|B)$ and simplifying yields the answer.

(c) When $x = 1/2$, the equilibrium always lies at $p = 1/2$. As x increases, the range of possible values leading to fixation of Bourgeois diminishes because as x increases, size better predicts victory, and Assessors, who use this information, do better than Bourgeois, who use a cue uncorrelated with fighting ability. Thus it stands to reason that animals should use correlated asymmetries when they can, but that uncorrelated asymmetries will be used when animals are approximately evenly matched. This is what Riechert and Hammerstein found in their experiments with spiders. ■

5. Errors in behavior

(a) Errors don't affect Doves so $V(D|D) = \frac{v}{2}$. When a Dove interacts with a Retaliator and no error occurs, they both behave like doves; if there is an error, the Retaliator acts like a Hawk. Thus

$$V(D|R) = (1-e)\frac{v}{2} = \frac{v}{2} - e\frac{v}{2}$$

$$V(R|D) = (1-e)\frac{v}{2} + ev = \frac{v}{2} + e\frac{v}{2}.$$

When two Retaliators interact and neither errs, both play as Doves. However, if either errs, the contest escalates and both play Hawk. Thus

$$V(R|R) = (1-e)^2 \frac{v}{2} + \left(1 - (1-e)^2\right) \frac{v-c}{2}.$$

Now, when e is small, $(1-e)^2 = 1 - 2e + e^2 \approx 1 - 2e$. Thus

$$V(R|R) \approx (1-2e)\frac{v}{2} + (1 - (1-2e))\frac{v-c}{2}$$

$$\approx (1-2e)\frac{v}{2} + e(v-c)$$

$$\approx \frac{v}{2} - ec.$$

(b) Retaliator can invade Dove when $V(R|D) = (1-e)\frac{v}{2} + ev = \frac{v}{2}(1+e) > \frac{v}{2}$, which is satisfied any time $e > 0$. Errors are good for Retaliators when they interact with Doves because they don't have to share.

(c) Dove can invade Retaliator if

$$\frac{v}{2} - ec < \frac{v}{2} - e\frac{v}{2}$$

$$-2ec < -e.$$

Assuming that $e > 0$, this inequality becomes

$$\frac{v}{2c} < 1.$$

(d) To find frequency of Retaliators at the mixed equilibrium, $\hat{p}$, set $W(R) = W(D)$ and solve $\hat{p}$:

$$\hat{p}W(R|R) + (1-\hat{p})W(R|D) = \hat{p}W(D|R) + (1-\hat{p})W(D|D)$$

$$\hat{p}\left(\frac{v}{2} - ec\right) + (1-\hat{p})\left(\frac{v}{2} + e\frac{v}{2}\right) = \hat{p}\left(\frac{v}{2} - e\frac{v}{2}\right) + (1-\hat{p})\frac{v}{2}$$

$$-\hat{p}(ec) + (1-\hat{p})e\frac{v}{2} = -\hat{p}e\frac{v}{2}.$$

Once again, if $e > 0$, we can solve

$$\hat{p} = \frac{v}{2c}.$$

(e) Hawks cannot invade this equilibrium when their fitness is lower than the fitness of Doves and Retaliators (who must have equal fitness):

$$W(D) = W(R) = w_0 + \hat{p}(1 - e)\frac{v}{2} + (1 - \hat{p})v.$$

The average fitness of rare Hawks will be

$$W(H) = w_0 + \hat{p}\frac{v - c}{2} + (1 - \hat{p})\frac{v}{2}.$$

Thus, Hawks won't be able to invade if

$$\hat{p}(1 - e)\frac{v}{2} + (1 - \hat{p})\frac{v}{2} > \hat{p}\frac{v - c}{2} + (1 - \hat{p})v$$

$$\hat{p}(-e)\frac{v}{2} > \hat{p}\frac{-c}{2} + (1 - \hat{p})\frac{v}{2}$$

$$0 > \hat{p}(-c - v + ev) + v.$$

Substituting in the expression for $\hat{p}$ from (d) yields

$$\frac{v}{2c}(-c - v + ev) + v < 0$$

$$(-c - v + ev) + 2c < 0$$

$$-v + ev + c < 0$$

$$e < 1 - \frac{c}{v}. \quad \blacksquare$$

6. Battle of the sexes

(a)   The simultaneous tree has the structure shown in Figure E.1.

(b)   In this case, the tree is as shown in Figure E.2. To compute the optimal strategy, start with Chris. If Pat chooses boxing, the the best choice for Chris is boxing; if Pat chooses opera, the best for Chris is opera. Now, look at it from Pat's perspective: If Pat chooses boxing, Chris will choose boxing, and Pat will get a payoff of 2. If Pat chooses the opera, Chris will choose the opera, and Pat will get 1. Thus the optimal choice for Pat is boxing. $\blacksquare$

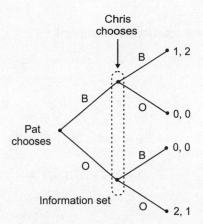

Figure E.1: The game tree for the simultaneous battle of the sexes game. The dotted line represents the information set for Chris and indicates that he (she?) doesn't know which node she (he?) is at.

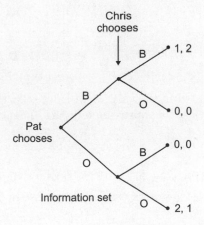

Figure E.2: The game tree for the sequential battle of the sexes game.

Chapter 3

1. Hawk-Dove among relatives

(a)   Using the definitions of the strategies and payoffs, we obtain

$$rV(D|C) + V(C|D) > rV(D|D) + V(C|D)$$
$$rb - c > r(0) + 0.$$

(b)   Now we let Hawk be strategy B and Dove be strategy A. This results in a new rule:

$$rV(H|D) + V(D|H) > rV(H|H) + V(H|H)$$
$$rv + 0 > r\frac{v-c}{2} + \frac{v-c}{2}$$
$$2rv > r(v-c) + v - c$$
$$r(2v - v + c) > v - c.$$

Then solve for r.

(c)   Let p be the frequency of Hawk. The haploid kin selection model says that the fitnesses of the strategies, respectively will be

$$W(H) = w_0 + \Pr(H|H)V(H|H) + \Pr(D|H)V(H|D)$$
$$= w_0 + [r + (1-r)p]V(H|H) + [(1-r)(1-p)]V(H|D)$$

and

$$W(D) = w_0 + \Pr(H|D)V(D|H) + \Pr(D|D)V(D|D)$$
$$= w_0 + [(1-r)p]V(D|H) + [r + (1-r)(1-p)]V(D|D).$$

We're only interested in the case when Dove is rare, so assume the frequency of Hawk, p, is approximately one. Then

$$W(H) = w_0 + V(H|H)$$

and

$$W(H) = w_0 + (1-r)V(D|H) + rV(D|D).$$

Doves (the "altruists" in this case) will increase in frequency when

$$W(D) > WH$$
$$(1-r)V(D|H) + rV(D|D) > V(H|H)$$
$$(1-r)0 + r\frac{v}{2} > \frac{v-c}{2}$$
$$r > \frac{v-c}{v} = 1 - \frac{c}{v}.$$

(d) Hamilton's rule results in the wrong answer because the Hawk-Dove payoffs are not additive. In the prisoner's dilemma, being helped always adds b to an animal's payoff and helping always subtracts c. In the Hawk-Dove game however, an animal's payoff against a Hawk or a Dove depends critically upon its own behavior. The payoffs interact in a nonlinear and nonadditive fashion, and so Hamilton's rule makes the wrong prediction, since its derivation depends upon additive fitness effects.

Note that kinship still matters here and can lead to self-sacrifice, but that the relationship is not described by Hamilton's rule. ■

2. Covariance assortment

First define two functions which relate y_i and h_i to p_i:

$$y_i = p_i \Pr(C|C) + (1 - p_i) \Pr(C|D)$$

and

$$h_i = p_i(1) + (1 - p_i)(0).$$

Read the first as: if $p_i = 1$, then the chance of being helped is $\Pr(C|C)$; if $p_i = 0$, then the chance is $\Pr(C|D)$. This is just definition, really. It says that the chance of being helped when the individual i has the allele for altruism is the probability of being helped when the individual has the allele for altruism. Pure tautology. The second is similar: if $p_i = 1$, then the chance of helping is one; if $p_i = 0$, then the chance of helping is zero.

Using the definition of a covariance, we can calculate $\text{cov}(p_i, y_i)$ and $\text{cov}(p_i, h_i)$. The expectation of $p_i y_i$ can be easily calculated by making a table like the one below:

Chance	p_i	y_i	$p_i y_i$
p	1	$\Pr(C|C)$	$\Pr(C|C)$
$1-p$	0	$\Pr(C|D)$	0

So the expectation is

$$E(p_i y_i) = p\Pr(C|C) + (1 - p)0 = p\Pr(C|C).$$

$E(p_i) = p$ and $E(y_i)$ is just

$$E(p_i y_i) = p\Pr(C|C) + (1 - p)\Pr(C|D).$$

Then we put these together to calculate the covariances:

$$\text{cov}(p_i, y_i) = p\Pr(C|C) - p\{p\Pr(C|C) + (1-p)\Pr(C|D)\}$$
$$= p(1-p)\{\Pr(C|C) - \Pr(C|D)\}.$$

The next covariance is calculated in the same way:

$$\text{cov}(p_i, h_i) = \text{E}(p_i h_i) - \text{E}(p_i)\,\text{E}(h_i).$$

A table will help again:

Chance	p_i	h_i	$p_i h_i$
p	1	1	1
$1-p$	0	0	0

Thus,

$$\text{E}(p_i h_i) = p(1) + (1-p)(0) = p.$$

So the expectation is just p. $\text{E}(h_i)$ is similarly p (as can be seen from the table above). Thus we have

$$\text{cov}(p_i, h_i) = p - p(p) = p(1-p).$$

Dividing the first covariance by the other results in the form of Hamilton's rule expressed with conditional probabilities of interaction:

$$\frac{\text{cov}(p_i, y_i)}{\text{cov}(p_i, h_i)}b > c$$
$$\frac{p(1-p)[\Pr(C|C) - \Pr(C|D)]}{p(1-p)}b > c$$
$$(\Pr(C|C) - \Pr(C|D))b > c. \quad \blacksquare$$

3. Whom do you help?

(a)   Inclusive fitness as a function of x_1, $I(x_1)$, is

$$I(x_1) = r_1(x_1)^{\frac{1}{2}} + r_2(R - x_1)^{\frac{1}{2}}.$$

(b)   The derivative of I is

$$\frac{dI(x)}{dx} = \frac{1}{2}r_1(x_1)^{-\frac{1}{2}} - \frac{1}{2}r_2(R - x_1)^{-\frac{1}{2}}.$$

Setting this equal to zero and solving for x_1 yields

$$\frac{1}{2}r_1(x_1)^{-\frac{1}{2}} - \frac{1}{2}r_2(R-x_1)^{-\frac{1}{2}} = 0$$

$$r_1(x_1)^{-\frac{1}{2}} = r_2(R-x_1)^{-\frac{1}{2}}$$

$$\left(\frac{r_1}{r_2}\right)(x_1)^{-\frac{1}{2}} = (R-x_1)^{-\frac{1}{2}}.$$

Raising both sides to the -2 power yields

$$\left(\frac{r_2}{r_1}\right)^2 x_1 = (R-x_1)$$

$$x_1 = \frac{R}{1 + \left(\frac{r_2}{r_1}\right)^2}.$$

Since $\frac{r_2}{r_1} < 1$, the denominator on the right-hand side is less than 2, which means that individual 1 gets more than individual 2.

(c) Now the derivative of inclusive fitness is given by

$$\frac{dI}{dx} = 2r_1x_1 - 2r_2(R-x_1) = -2r_2R + 2(r_1 + r_2)x_1.$$

Setting this equation equal to zero yields

$$r_2R = (r_1 + r_2)x_1.$$

Dividing by r_2 and solving yields the desired expression. Since $\frac{r_1}{r_2} > 1$, the denominator is greater than 2, and therefore individual 1 gets less. The derivative is negative for values of x_1 less than this value and positive for larger values. That means that this value of x_1 is a local minimum, and since there is only one such value the maximum of the function must be at either 0 or R. Since $r_1R^2 > r_2R^2$, the maximum value is $x_1 = R$. The reason is that the effects of giving are subject to increasing returns to scale—doubling the endowment more than doubles the fitness effect—so the best thing is to give everything to the most closely related individual. When there are increasing returns to scale, Hamilton's rule does not predict that resources are (or help more generally is) doled out in proportion to relatedness. ∎

4. Frequency-dependent assortment

(a) From the text we know that the altruistic type will increase if

$$\{\Pr(A|A) - \Pr(A|S)\}b > c.$$

Thus we need to compute $\Pr(A|A)$ and $\Pr(A|S)$. From Bayes' law,

$$\Pr(A|A) = \frac{\Pr(A, A)}{\Pr(A)} = \frac{\Pr(A, A)}{p} = \frac{p(1 + a)}{1 + ap}.$$

Similarly

$$\Pr(A|S) = \frac{\Pr(A, S)}{\Pr(S)} = \frac{\Pr(A, S)}{1 - p} = p\left(\frac{1}{1 + ap} + \frac{1}{1 + a(1 - p)}\right).$$

Thus

$$\Pr(A|A) - \Pr(A|S) = p\left(\frac{1 + a}{1 + ap} - \frac{1}{1 + ap} - \frac{1}{1 + a(1 - p)}\right)$$

$$= p\left(\frac{a}{1 + ap} - \frac{1}{1 + a(1 - p)}\right).$$

Now we can answer the question. When $p \to 0$, $\Pr(A|A) - \Pr(A|S) \to 0$, and therefore altruism cannot increase.

(b) When $p \to 1$, $\Pr(A|A) - \Pr(A|S) \to \frac{-1}{1+a}$, and altruism really can't increase.

(c) To show that p increases for large enough a and intermediate values of p choose $p = 0.5$. Then

$$\Pr(A|A) - \Pr(A|S) = 0.5\left(\frac{1 + a}{1 + ap} - \frac{1}{1 + 0.5a} - \frac{1}{1 + 0.5(1 - p)}\right)$$

$$= \frac{a - 1}{2 + a} \to 1 \text{ as } a \to \infty.$$

Thus for large enough a, p will increase as long as $b > c$. Unlike the case of interaction between relatives, the degree of assortment here is frequency dependent. When altruists are rare, interaction is approximately random, and there is not enough assortment for altruism to increase. However, as altruists become more common, assortment increases, and therefore altruism can evolve. This kind of phenotypic clustering has proven to be quite common in ecology, and may provide a mechanism for the evolution of altruism without genealogical relatedness. However, if altruism is very important, selection can shape patterns of dispersal so as to negate this effect. ■

Chapter 4

1. Tiny basins

Let p be the frequency of TFT. Find the internal equilibrium as explained in Chapter 2, by setting $W(TFT) = W(ALLD)$, $p = \hat{p}$, and solve for $\hat{p}$. The two fitness expressions in this case are:

$$W(ALLD) = w_0 + pb + (1-p)(0)$$
$$W(TFT) = w_0 + p\frac{b-c}{1-w} - (1-p)c.$$

Setting $p = \hat{p}$ and solving:

$$W(TFT) = W(ALLD)$$

$$\hat{p}\frac{b-c}{1-w} - (1-\hat{p})c = \hat{p}b$$

$$\hat{p}\left(\frac{b-c}{1-w} + c - b\right) = c$$

$$\hat{p}\left(\frac{b-c}{1-w} - (b-c)\right) = c$$

$$\hat{p}\left((b-c)\frac{w}{1-w}\right) = c.$$

One last step to isolate $\hat{p}$ yields the answer. We know that this unique internal equilibrium is unstable, because both pure equilibria are stable.

As w increases, $\hat{p}$ approaches zero. This means the basin of attraction of ALLD diminishes as relationships last longer. Suppose $b/c = 2$. Then $\hat{p} = (1-w)/w$. For $w = 0.5$, $\hat{p} = 1$. But for $w = 0.9$ (10 interactions on average), $\hat{p} \approx 0.11$. If b/c is large and w is even moderately large, the domain of attraction for ALLD is vanishingly small. ■

2. Even bad guys are vulnerable

Consider when ALLD is common, such that p is the frequency of ALLD, and $p \approx 1$. Let q and $r = 1 - p - q$ be the frequencies of rare STFT and TF2T individuals, respectively. The fitness of ALLD is:

$$W(ALLD) = w_0 + p(0) + q(0) + r(2b).$$

The fitness expressions for STFT and TF2T are:

$$W(STFT) = w_0 + p(0) + q(0) + r\left(b + \frac{w}{1-w}(b-c)\right)$$

$$W(TF2T) = w_0 + p(-2c) + q\left(-c + \frac{w}{1-w}(b-c)\right) + r\frac{b-c}{1-w}.$$

TF2T cannot invade, because its fitness is approximately $w_0 - 2c$. However, ALLD and STFT have equivalent fitness expressions up to the term including r. Then STFT will invade provided:

$$W(STFT) > W(ALLD)$$

$$\frac{w}{1-w}(b-c) > b.$$

We could muck around with the above further, solving for w, but it is clear already that if the left—the benefit of cooperation after the first turn—exceeds the right side—the benefit of being helped without providing help in return on the first turn only—then STFT will invade a population of ALLD. The reason is that rare TF2T individuals allow STFT to occasionally end up in long-lasting partnerships, where ALLD never does. ∎

3. More tolerance

(a) The payoffs are

$$V(TF2T|ALLD) = -c - wc$$

and

$$V(TF2T|TF2T) = \frac{b-c}{1-w}.$$

If p is the frequency of TF2T, the fitness becomes

$$W(TF2T) = w_0 + p\frac{b-c}{1-w} + (1-p)(-c-wc).$$

The payoff to ALLD against TF2T is

$$V(ALLD|TF2T) = b + wb.$$

So TF2T resists invasion by ALLD when

$$V(TF2T|TF2T) > V(ALLD|TF2T)$$
$$\frac{b - c}{1 - w} > b + wb$$
$$b - c > b(1 + w)(1 - w)$$
$$b - c > b(1 - w^2)$$
$$w^2 b > c.$$

This means it is easier for ALLD to invade TF2T than TFT. The condition for TFT to be an ESS against ALLD is

$$wb > c.$$

For any given w, this is easier to satisfy than the condition you just derived. This suggests more-tolerant strategies will be more easily invaded by cheaters. However, if there are rare errors, as we saw in the text, then more-tolerant strategies will avoid sequences of retaliation against themselves, and this might alter the conclusions you derived above.

(c) Prove this by writing a general payoff to ALLD against TFnT:

$$V(ALLD|TFnT) = b + wb + w^2 b + \ldots + w^{n-1} b$$
$$= b(1 + w + w^2 + \ldots + w^{n-1}).$$

TFnT is an ESS when

$$\frac{b - c}{1 - w} > b(1 + w + w^2 + \ldots + w^{n-1})$$
$$b - c > b(1 - w)(1 + w + w^2 + \ldots + w^{n-1})$$
$$b - c > b(1 - w + w - w^2 + w^2 - \ldots - w^{n-1} + w^{n-1} - w^n).$$

Each second term cancels each following first term, so we get

$$b - c > b(1 - w^n).$$

This immediately simplifies to the answer. ∎

4. Doing the dishes again, and again

(a) WFT alternates washing the dishes with itself. To calculate this payoff, notice that this implies two infinite repeating series: one when the focal individual washes first, and one when the focal individual washes second.

$$V(WFT|WFT) =$$

$$\frac{1}{2}\left[(b-k) + wb + w^2(b-k) + w^3b + \ldots\right]$$

$$+ \frac{1}{2}\left[b + w(b-k) + w^2b + w^3(b-k) + \ldots\right]$$

$$= \frac{1}{2}\left[b + wb + w^2b + \ldots\right] - \frac{1}{2}\left[k + w^2k + w^4k + \ldots\right]$$

$$+ \frac{1}{2}\left[b + wb + w^2b + \ldots\right] - \frac{1}{2}\left[wk + w^3k + w^5 + \ldots\right]$$

$$= b(1 + w + w^2 + w^3 + \ldots)$$

$$- \frac{k}{2}(1 + w + w^2 + w^3 + \ldots)$$

$$= \frac{b - k/2}{1 - w}.$$

Since each player is equally likely to be first or second, payoffs average out as if they weren't taking turns at all, but instead also both washing. This is only true because we assumed that washing together halves the cost of washing alone per player.

(b) The payoff to NW against WFT is

$$V(NW|WFT) = b.$$

This is because NW refuses to wash, so WFT washes on the first turn. WFT then imitates its opponent on each subsequent turn, meaning neither player ever washes again. WFT is an ESS against NW when

$$V(WFT|WFT) > V(NW|WFT)$$

$$\frac{b - \frac{k}{2}}{1 - w} > b$$

$$b - \frac{k}{2} > b - wb$$

$$wb > \frac{k}{2}.$$

This condition is very similar to Hamilton's rule. w expresses the nonrandom persistence of groups of cooperators. $\frac{k}{2}$ is the cost of cooperating (washing) when the other player is already washing. Thus this substitutes for c in the altruism model. ∎

5. Sequential reciprocity

First compute the fitness of each type when interacting with itself and the other type. We start with TFT. Since an individual is equally likely to be donor and recipient on the on the first turn,

$$V(TFT|TFT) = \overbrace{\frac{1}{2}\left(b - wc + w^2b - w^3c + \ldots\right)}^{\text{recipient first}}$$

$$+ \overbrace{\frac{1}{2}\left(-c + wb - w^2c + w^3b + \ldots\right)}^{\text{donor first}}$$

$$= \frac{1}{2}\left(b - c + w(b-c) + w^2(b-c) + \ldots\right)$$

$$= \frac{1}{2}\frac{b-c}{1-w}.$$

Similarly

$$V(TFT|ALLD) = \frac{1}{2}(0) + \frac{1}{2}(-c).$$

The same procedure allows us to calculate the payoffs to ALLD:

$$V(ALLD|ALLD) = 0$$

$$V(ALLD|TFT) = \frac{1}{2}(0) + \frac{1}{2}(b).$$

Notice that these payoffs are exactly the same as in the Axelrod-Hamilton model (except that b and c are multiplied by $1/2$). But all of the equilibrium and invasion properties of the model depend only on the ratio of b and c. Thus, the sequential model has identical equilibrium and invasion properties as the simultaneous model. ∎

6. Tolerance and n-person reciprocity

When ALLD is rare, the average fitness of both types is determined by groups in which all of the other $n - 1$ individuals in the group use the strategy T_j. Then from the text we know that

$$V(T_j|n-1) = \frac{b-c}{1-w}.$$

The fitness of ALLD in such groups is

$$V(ALLD|n-1) = (n-1)\left(\frac{b}{n}\right)(1+w+w^2+\ldots) = \frac{\left(1-\frac{1}{n}\right)b}{1-w}.$$

Thus, $V(T_j|n-1) > V(ALLD|n-1)$ requires that $c - \frac{b}{n} > 0$. However, this is the condition for cooperation to increase when $w = 0$, that is, when cooperation is individually beneficial. ∎

7. Nice guys finish last

(a) The payoff depends critically upon whether the pair will persist or end after one round. So the payoff expression has a term for each event, weighted by v:

$$V(TFT|TFT) = \underbrace{v\left(\frac{b-c}{1-w}\right)}_{\text{pair persists}} + \underbrace{(1-v)(b-c)}_{\text{pair doesn't persist}}$$

$$= (b-c)\left(\frac{v}{1-w} + \frac{(1-v)(1-w)}{1-w}\right)$$

$$= (b-c)\left(\frac{v+1-w-v+wv}{1-w}\right)$$

$$= (b-c)\left(\frac{1-w(1-v)}{1-w}\right).$$

(b) TFT is an ESS against ALLD when

$$V(TFT|TFT) > V(ALLD|TFT)$$

$$(b-c)\left(\frac{1-w(1-v)}{1-w}\right) > b$$

$$(b-c)(1-w(1-v)) > b - wb$$

$$wb - (wb - wc)(1-v) > c$$

$$vw(b-c) + wc > c$$

$$v\frac{w(b-c)}{1-w} > c$$

(c) HTFT against itself misses cooperation on the first round but cooperates always thereafter. Its payoff against itself must be

$$V(TFT|TFT) = v\frac{w(b-c)}{1-w} + (1-v)(0).$$

To see this, note that the expanded payoff is:

$$V(TFT|TFT) = v\left[w(b-c) + w^2(b-c) + w^3(b-c) + \ldots\right] + v(0)$$
$$= vw(b-c)\left[1 + w + w^2 + w^3 + \ldots\right].$$

The infinite series is the one most familiar to us, leading directly to the concise payoff expression.

ALLD against HTFT gains

$$V(ALLD|TFT) = v(0 + wb) + (1-v)(0)$$
$$= vwb.$$

HTFT defects on the first turn, providing no benefit to ALLD. If interactions do not end (v of the time), it gains b from HTFT's initial cooperation, provided the pair continues (w of the time).

HTFT will be ESS against ALLD when

$$v\frac{w(b-c)}{1-w} > vwb$$
$$vwb - vwc > vwb - vw^2b$$
$$wb > c.$$

HTFT more easily resists invasion by ALLD. You can see this by noting that

$$wb > v\frac{w(b-c)}{1-w}$$

whenever

$$v < \frac{1-w}{1-\frac{c}{b}}.$$

Thus as long as v, the chance that interactions ever persist beyond the first round, is small enough, HTFT is better able to resist invasion because it waits to see if interactions continue before attempting to cooperate. ∎

8. n-person stag hunting

(a) When S is common, it always finds itself in a group with $n-1$ other S individuals. Its payoff is therefore

$$V(S|n-1) = \frac{\overbrace{n-1}^{\text{others}} + \overbrace{1}^{\text{self}}}{n} s = s.$$

The payoff to H is always h, regardless of group composition. The condition for S to be an ESS is then $s > h$, which is always satisfied.

When H is common, a rare S finds itself in a group of $n-1$ H's. It's payoff is then $V(S|0) = \frac{1}{n}s$. The condition for H to be an ESS is therefore

$$h > \frac{s}{n},$$

or

$$n > \frac{s}{h}.$$

The group size must exceed the ratio of s to h for H to be an ESS. Thus H is more likely to be an ESS as group size increases.

(b) When both conditions above are satisfied, there is an internal unstable equilibrium. Let p be the frequency of S. The payoff to an S in a group of x other S individuals is

$$V(S|n-1) = \frac{x+1}{n}s.$$

The fitness of an S individual across all possible groups is

$$\begin{aligned}
W(S) &= w_0 + \sum_{x=0}^{n-1} k(x)V(S|x) \\
&= w_0 + \sum_{x=0}^{n-1} k(x)\frac{x+1}{n}s
\end{aligned}$$

First factor constants out of the sum, then take the expectation:

$$\begin{aligned}
W(S) &= w_0 + \frac{s}{n}\sum_{x=0}^{n-1} k(x)(x+1) \\
&= w_0 + \frac{s}{n}\left(\sum_{x=0}^{n-1} k(x)x + \sum_{x=0}^{n-1} k(x)\right) \\
&= w_0 + \frac{s}{n}\left(\sum_{x=0}^{n-1} k(x)x + 1\right).
\end{aligned}$$

The remaining sum is just the mean of the binomial distribution with $n - 1$ trials, so

$$W(S) = w_0 + \frac{s}{n}[p(n-1) + 1].$$

Now set this equal to $W(H)$ and solve for $\hat{p}$:

$$\frac{s}{n}[\hat{p}(n-1) + 1] = h,$$

or

$$\hat{p} = \frac{n\dfrac{h}{s} - 1}{n - 1}.$$

As the ratio of h to s gets smaller, the numerator gets smaller, increasing the domain of attraction for S. Note that as n increases, $\hat{p}$ approaches one, shrinking the domain of attraction for S. This is because we assumed that, as groups get large, it takes more and more hunters to ensure capture of a stag. ∎

Chapter 5

1. It is always OK to say nothing

(a) Assume that $a/d < r < b/d$, so that signaling is favored only if it is costly. First figure out the inclusive fitness of the common-type donor. We focus on the donors because if AG is common, then a beneficiary who signals, at cost c, will never do better than one who does not. Thus, NS is stable, and we need only consider whether AG is stable as well. Common-type AG donors always give and so have fitness

$$V_D(AG|NS) = (1 - d),$$

and beneficiaries always receive and so have fitness

$$V_B(NS|AG) = 1.$$

Thus the inclusive fitness of the common type (AG/NS) is $1 - d + r(1)$. Now figure out the inclusive fitness of the invading R or NG donor. Since beneficiaries don't signal, the fitness of R donors and NG donors is the same:

$$V_D(R|NS) = V_D(NG|NS) = (1 - p) + p = 1.$$

Similarly the fitness of NS beneficiaries in both cases is

$$V_B(NS|R) = V_B(NS|NG) = (1-p)(1-b) + p(1-a).$$

The inclusive fitness of both invading types is therefore $1 + r((1-p)(1-b) + p(1-a))$. Thus, AG can resist invasion by R and NG whenever

$$1 - d + r > 1 + r((1-p)(1-b) + p(1-a))$$
$$d < r((1-p)b + pa).$$

If the cost to the donor of transferring the resource is small enough, it will always pay to give.

(b) When NG/NS is common, donors have inclusive fitness $1 + r((1-p)b + pa)$. R has the same fitness as a donor, but AG has inclusive fitness $(1-d) + r$. Thus NG can resist invasion by AG as long as $d > r((1-p)b + pa)$. R has no selective advantage but can drift into the population. ∎

2. Sometimes it's better to say nothing at all

(a) In the signaling population, donors have inclusive fitness

$$\underbrace{1 - pd}_{V_D(R|R)} + \underbrace{r\left((1-p)(1-b) + p(1-\hat{c})\right)}_{V_R(R|R)},$$

where $\hat{c} = b - rd$ is the smallest value that allows honest signaling to persist. Then the inclusive fitness of donors is

$$1 - pd + r\left((1-p)(1-b) + p(1-b+rd)\right).$$

In the nonsignaling population, donors have inclusive fitness $1 - d + r$ (see the solution to Problem 1). Thus the donors in the nonsignaling population have higher fitness if

$$1 - d + r > 1 - pd + r\left((1-p)(1-b) + p(1-b+rd)\right).$$

Simplifying gives

$$0 > d(1-p) - rb + r^2 pd,$$

which yields the desired condition

$$b > \frac{d}{r}(1-p) + prd.$$

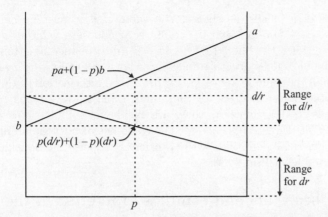

Figure E.3: A graphical proof that there are conditions in which the nonsignaling equilibrium has higher fitness than the signaling equilibrium.

(b) In the signaling population, beneficiaries have inclusive fitness

$$\underbrace{r\left((1-p)(1) + p(1-d)\right)}_{V_D(R|R)} + \underbrace{(1-p)(1-b) + p(1-\hat{c})}_{V_B(R|R)}.$$

In the nonsignaling population, beneficiaries have inclusive fitness $r(1-d) + 1$. Thus the beneficiaries in the nonsignaling population have higher fitness if

$$r(1-d)+1 > r\left((1-p)+p(1-d)\right)+(1-p)(1-b)+p\left(1-(b-rd)\right).$$

Simplifying gives

$$0 > rd(1-p) - (1-p)b - p(b-rd).$$

This becomes

$$b > rd,$$

which is always true if signals are necessary.

(c) For costly signaling to be stable, $a > d/r > b$. For AG/NS to be an equilibrium, $d/r < pa+(1-p)b$. This places an upper bound on d/r. Figure E.3 plots $pa + (1 - p)b$ as a function of possible values of p. Thus, as shown, the value of d/r must lie between $d/r < pa + (1-p)b$ and $d/r > b$. Now choose a value of d/r as shown. Then the line $p'(d/r) + (1-p')(dr)$ (defined as a function of possible values of p, p') must intersect the left axis at b and go through the point $p(d/r) + (1-p)(dr)$. The intersection of this line with the right axis gives the maximum value of dr. As long as this value is greater than zero, there are parameter combinations which satisfy the three conditions for the existence of both signaling and nonsignaling equilibrium with transfers and cause the nonsignalers to have higher fitness. ■

3. Cheap talk and conflicts of interest in the battle of the sexes game

(a)   When two D's are paired, they always rendezvous. The payoff of the focal individual is g with probability γ and b with probability $1 - \gamma$. Thus

$$V(D|D) = \gamma g + (1-\gamma)b.$$

when D is common, are W's never rendezvous, so have $V(W|D) = 0$. A rare P individual can rendezvous only when it prefers dry. Thus

$$V(P|D) = \gamma g.$$

Because $V(D|D) > V(P|D) > V(W|D)$, D is an ESS.

(b)   A pair of honest signalers always rendezvous. The male always gets his preferred site. The female gets her preferred site when they agree and the less preferred site when they don't. Since individuals are equally likely to be males and females,

$$V(H|H) = \frac{1}{2}g + \frac{1}{2}\left(\gamma^2 g + (1-\gamma)^2 g + 2\gamma(1-\gamma)b\right).$$

Simplifying yields

$$V(H|H) = \{1 - \gamma(1-\gamma)\}g + \gamma(1-\gamma)b.$$

Rare invading D's don't rendezvous any time the other individual prefers W, and thus can't invade:

$$V(D|H) = \gamma\{\gamma g + (1-\gamma)b\} + (1-\gamma)0.$$

Above, γ of the time, the other individual prefers the dry site. When this is true, γ of the time the focal individual also prefers

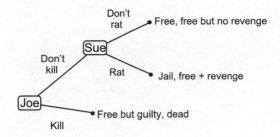

Figure E.4: The game tree for Chapter 5, Problem 4, part a.

it and gets g, otherwise it prefers wet and gets b. The rest of the time, the other individual prefers wet, and so either signals wet and goes there (male) or fails to find a signal and goes there anyway (female).

$V(H|H) > V(D|H)$ as long as $\gamma < 1$. That is, if there is any possibility of a conflict of interest, signaling is an ESS against choosing dry. ∎

4. External commitment

(a) The game tree is given in Figure E.4. Start at the end of the game. If Joe does not kill Sue, she prefers to rat Joe out in revenge for the kidnapping. Notice the crucial role that a small preference for revenge plays. It will be some trouble for Sue to accuse Joe, participate in the trial, etc. If she did not get any utility from seeing Joe go to jail, she wouldn't rat, even though she could. Now consider Joe's decision. If he kills Sue, he will be alive, free, and feel a bit guilty. If he doesn't kill her, she will rat, and he will go to jail. As long as a bit of guilt is better than jail, he chooses to kill his victim.

(b) The game tree is given in Figure E.5. Once again, start at the end of the game. If Sue has committed the crime and then ratted Joe out, he will do the same. Again notice the key role played by a preference for vengeance. Now consider Sue's decisions. If she agrees to commit the crime, and she rats Joe out, she knows she will go to jail. If she doesn't, she forgoes revenge but avoids jail. So as long as Sue prefers freedom to the satisfaction of revenge she

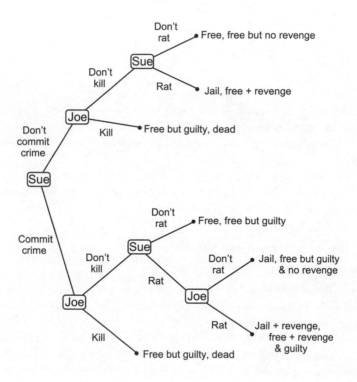

Figure E.5: The game tree for Chapter 5, Problem 4, part b.

won't rat. This means that if Joe doesn't kill he goes free, while if he does, he goes free but suffers pangs of guilt. So, he doesn't kill her. ■

5. Imperfect individual learning

The recursion for the mean value of q is

$$\bar{q} = (1 - u)E\{(1 - p)(1 - e) + pq\}$$

$$\bar{q} = \frac{(1 - p)(1 - u)(1 - e)}{1 - p(1 - u)}.$$

We can then compute the fitness of social learners at this equilibrium:

$$\hat{W}(S) = w_0 + b\frac{(1 - p)(1 - u)(1 - e)}{1 - p(1 - u)}.$$

To find the equilibrium, set

$$\hat{W}(S) = W(I)$$

$$b(1 - e)\frac{(1 - \hat{p})(1 - u)}{1 - \hat{p}(1 - u)} = b(1 - e) - c.$$

Solving for $\hat{p}$ yields the desired result. (Simply substitute $b(1 - e)$ for b in the derivation given in Chapter 5.) ■

Chapter 6

1. When the strength of selection among groups varies

(a) The payoffs in this game are

$$V(H) = h$$

$$V(S) = ps + (1 - p)0 = ps.$$

H is an ESS because $h > 0$. S is an ESS because $s > h$. There is one internal equilibrium at $\hat{p} = \frac{h}{s}$. This equilibrium must be unstable because both of the pure equilibria are stable. As h increases relative to s, the domain of attraction for S shrinks.

(b) We can use a table to simplify computing the covariance across groups. We can cut some corners if we recall that $\operatorname{cov}(x, y) = \operatorname{var}(x)\beta(y, x)$:

p_g	Pr	$\operatorname{var}(p_{ig})$
0	$(1-p)^2$	0
1/2	$2p(1-p)$	$\operatorname{E}(p_{ig}^2) - \operatorname{E}(p_{ig})^2 = \frac{1}{4}$
1	p^2	0

Note that the only groups in which there is any variance in genotype are those with one S and one H (seems obvious enough). Thus these are the only groups which contribute to the expected covariance. Thus we just need now the regression of individual fitness on genotype in mixed pairs. The individual with $p_i = 0$ has fitness h and the individual with $p_i = 1$ has fitness 0. Thus the slope is $(0 - h)/(1 - 0) = -h$. All together, the expected covariance of individual fitness and individual genotype is

$$\operatorname{E}\{\operatorname{cov}(w_{ig}, p_{ig})\} = 2p(1 - p)\frac{1}{4}(-h).$$

The covariance between group fitness and group genotype is also made easier by a table:

p_g	Pr	w_g	$p_g w_g$
0	$(1-p)^2$	h	0
1/2	$2p(1-p)$	$h/2$	$h/4$
1	p^2	s	s

Thus the covariance is

$$\operatorname{cov}(w_g, p_g) = \operatorname{E}(p_g w_g) - \operatorname{E}(p_g)\operatorname{E}(w_g)$$

$$= p^2 s + 2p(1-p)\frac{h}{4}$$

$$+ (1-p)^2 0 - p\left(p^2 s + 2p(1-p)\frac{h}{2} + (1-p)^2 h\right)$$

$$= p(1-p)\left(ps - \frac{h}{2}\right).$$

(c) Selection within groups is always negative, acting against S. This is because in mixed pairs, H always does better than S. Selection among groups favors S provided

$$ps - \frac{h}{2} > 0$$

$$p > \frac{h}{2s}.$$

If S is common enough, selection among groups favors it. When it is rare, however, selection among groups favors H. The reason is that mean fitness in a pair is a nonlinear function of mean genotype in the pair. When $p \approx 0$, the regression of w_g on p_g is negative because groups with only one S individual have lower mean fitness than those with no S individuals. It isn't until p is large that a sufficient number of groups with two S individuals exist and tip the regression line upwards. ■

2. Covariance reciprocity

We know from the chapter that the variance among pairs is

$$\text{var}(p_g) = \frac{1}{2}p(1 - p)(1 + r),$$

and the expected variance within pairs is

$$\text{E}\{\text{var}(p_{ig})\} = \frac{1}{2}p(1 - p)(1 - r).$$

Now we just need the regression coefficients. The regression of individual fitness on individual genotype (in groups in which one individual is TFT and the other is ALLD) is

$$\beta(w_{ig}, p_{ig}) = \frac{-c - b}{1 - 0} = -b - c,$$

just like in the kin-selection case. This makes sense because the IPD is essentially a one-shot PD between an ALLD and a TFT player.

The regression of group fitness on group genotype is more complicated. Note that group fitness is not a linear function of group genotype in this case. Thus the regression will depend upon p and r because as these values change, the frequency of pairs with different genotypes changes, altering the regression coefficient. The straightforward way to compute the regression is to compute the covariance between p_g and w_g and factor out the variance, which we know to be $\frac{1}{2}p(1 - p)(1 + r)$. Using our familiar table, we have the following

Group	Frequency	p_g	w_g	$w_g p_g$
ALLD, ALLD	$(1-p)(r+(1-r)(1-p))$	0	0	0
ALLD, TFT	$2p(1-p)(1-r)$	0.5	$\frac{b-c}{2}$	$\frac{b-c}{4}$
TFT, TFT	$p(r+(1-r)p)$	1	$\frac{b-c}{1-w}$	$\frac{b-c}{1-w}$

Then we can compute the covariance:

$$\operatorname{cov}(w_g, p_g) = \operatorname{E}(w_g p_g) - \operatorname{E}(w_g)\operatorname{E}(p_g)$$

$$= \left\{ 2p(1-p)(1-r)\tfrac{b-c}{4} + p(r+(1-r)p)\tfrac{b-c}{1-w} \right\}$$

$$- \left\{ 2p(1-p)(1-r)\tfrac{b-c}{2} + p(r+(1-r)p)\tfrac{b-c}{1-w} \right\}\{p\}.$$

After some algebra we obtain

$$\operatorname{cov}(w_g, p_g) = \underbrace{\frac{1}{2}p(1-p)(1+r)}_{\operatorname{var}(p_g)} \underbrace{\frac{b-c}{1-w}\left(1 - w\frac{(1-2p)(1-r)}{1+r}\right)}_{\beta(w_g,p_g)}.$$

The regression within groups is always negative. Thus selection within groups favors ALLD. The regression among groups is positive, and its magnitude depends upon the values of p and r because these values determine the frequency of the different groups. When $r = 0$ and $p \approx 0$, we get the original invasion criterion for TFT in the IPD:

$$\bar{w}\Delta p \approx \frac{1}{2}p(1-p)\left((-b-c) + \frac{b-c}{1-w}(1-w)\right)$$

$$\approx p(1-p)(-c).$$

Thus TFT will not invade ALLD when $r = 0$ unless altruism is individually beneficial. Now, with $r > 0$,

$$\bar{w}\Delta p \approx \frac{1}{2}p(1-p)\left((1-r)(-b-c) + (1+r)\frac{b-c}{1-w}\left(1 - w\tfrac{1-r}{1+r}\right)\right).$$

Thus TFT invades provided

$$(1-r)(-b-c) + (1+r)\frac{b-c}{1-w}\left(1 - w\frac{1-r}{1+r}\right) > 0$$

$$r > \frac{1-w}{b/c - w}.$$

This is the same expression we derived in Chapter 4 for TFT to invade under kin selection. ∎

3. Local dispersal and the Hamilton-May model

As in Chapter 6, we compute the probability that a mutant dis-perser is successful at a common site, $\Pr(M|C)$. In this case

$$\Pr(M|C) = \frac{\overbrace{\dfrac{(v+\delta)pk}{2}}^{\text{mutants reaching site}}}{\underbrace{(1-v)k}_{\text{home-bodies}} + \underbrace{\dfrac{(v+\delta)pk}{2} + \dfrac{vpk}{2}}_{\text{all dispersers reaching site}}}.$$

To get $\mathrm{E}(\#M|C)$, multiply the above by 2 because the mutant only sends dispersers to the two neighboring sites in this model, not the other $n-1$ sites, as in the chapter. Now compute $\Pr(M|M)$:

$$\Pr(M|M) = \frac{(1-v-\delta)k}{(1-v-\delta)k + \dfrac{vpk}{2} + \dfrac{vpk}{2}}.$$

As in the original model, there is only one of these sites, so $\mathrm{E}(\#M|M)=\Pr(M|M)$. The equilibrium fraction of dispersers, $v^\star$, satisfies

$$\mathrm{E}(\#M|C) + \mathrm{E}(\#M|M) < 1.$$

Substituting in the expressions for $\mathrm{E}(\#M|C)$ and $\mathrm{E}(\#M|M)$ and simplifying yields the answer:

$$v^\star = \frac{2}{4-p} = \frac{1}{2 - \dfrac{p}{2}}.$$

There are fewer dispersers at equilibrium in this model at all sur-vival rates $p > 0$. At $p \approx 0$, selection favors sending almost half of the offspring to almost certain death, as in the original model. However, if $p \approx 1$, such that dispersers survive in all likelihood, then the ESS fraction of dispersers is 2/3 in this model, whereas it was almost 1 in the original model. It is still true that at most one offspring can succeed at the home site, but now the offspring do not disperse as far and end up competing more with one another at the two dispersal sites. In the original model, the number of sites was very large and dispersal so wide that dispersers essen-tially never competed with other dispersers from the same home site. Increased competition among dispersers results in fewer dis-persers at equilibrium. This result implies that there is an optimal distance of dispersal, as W. D. Hamilton speculated in several of his early papers on the evolution of social behavior. ∎

4. Sex-specific selection on genes on the X chromosome

Begin by writing the difference equation based on the Price equation. The change in the frequency of the G allele is given by

$$\bar{w}\Delta p = \text{cov}(w_i, p_i) = \frac{2}{3}\text{cov}_f(w_i, p_i) + \frac{1}{3}\text{cov}_m(w_i, p_i),$$

where cov_f is the covariance for females only and cov_m is for males only. Since males hold only 1/3 of the total number of alleles at this locus, they contribute only 1/3 to the replicator dynamic. Think of it this way: if we select an X-chromosome at random in the population, what is the chance it is in a male body?

Now compute the covariance for females. A handy table helps us do this:

Genotype	Frequency	p_i	w_i	$w_i p_i$
NN	$(1-p)^2$	0	$1-d$	0
GN	$2p(1-p)$	0.5	$1-d/2$	$\frac{1}{2}(1-d/2)$
GG	p^2	1	1	1

Thus

$$\begin{aligned}
\text{cov}_f(w_i, p_i) &= \text{E}(w_i p_i) - \text{E}(w_i)\,\text{E}(p_i) \\
&= p^2(1) + p(1-p)(1-d/2) \\
&\quad - \left(p^2 + 2p(1-p)(1-d/2) + (1-p)^2(1-d)\right)p \\
&= \frac{1}{2}p(1-p)d,
\end{aligned}$$

which is just the variance for a diploid organism and the regression (d) of fitness on allele frequency in this case. The covariance for males is done in the same way:

Genotype	Frequency	p_i	w_i	$w_i p_i$
N	$1-p$	0	1	0
G	p	1	$1-h$	$1-h$

This gives

$$\begin{aligned}
\text{cov}_m(w_i, p_i) &= p(1-h) + (1-p) - \left(p(1-h) + (1-p)\right)p \\
&= p(1-p)(-h).
\end{aligned}$$

Combining the two give

$$\text{cov}(w_i, p_i) = \frac{1}{3}p(1-p)(d-h).$$

So the G allele increases provided $d > h$. ∎

Chapter 7

1. Big caterpillars and female wasps

(a) Let r_1 be the fraction of males produced on big caterpillars and r_0 be the proportion of males produced on small caterpillars. Then from the rule given in the chapter, producing all females on big caterpillars will be evolutionarily stable only if:

$$\frac{\partial(fm)}{\partial r_1} < 0 \Rightarrow 2 < \frac{f}{m}.$$

But

$$\frac{f}{m} = \frac{2g + 0(1-g)}{g(0) + (1-g)} = \frac{2g}{(1-g)}.$$

Therefore all females on big caterpillars is stable if

$$2 > \frac{2g}{(1-g)} \Rightarrow g < 1 - g \Rightarrow g < \frac{1}{2}.$$

(b) Then producing all males on small caterpillars will be evolutionarily stable if

$$\frac{\partial(fm)}{\partial r_0} > 0 \Rightarrow 1 > \frac{f}{m},$$

and therefore all females on big caterpillars is stable if

$$1 < \frac{2g}{(1-g)} \Rightarrow 2g > 1 - g \Rightarrow g > \frac{1}{3}.$$

(c) Genes increasing the production of females will continue to invade until

$$\frac{\partial(fm)}{\partial r_0} = 0 \Rightarrow 1 = \frac{f}{m}.$$

Assume that females still produce all females on big caterpillars but now produce a proportion r_0 of males on small caterpillars. We then find the value of r_0 that makes $f/m = 1$. Thus

$$\frac{f}{m} = \frac{2g + (1-r_0)(1-g)}{(1-r_0)(1-g)} = 1.$$

Solving for g yields

$$2g + (1-g)(1-r_0) = (1-g)r_0$$
$$2g + (1-g) = 2(1-g)r_0$$
$$r_0 = \frac{1}{2}\left(\frac{1+g}{1-g}\right)$$

For $g = 1/4$ this becomes $5/6$. ∎

2. Inbreeding and local mate competition

When sib matings are present, the fitness of a rare invading female producing a fraction $1 - \hat{r}$ daughters given that most females produce a fraction $1 - r$ daughters is

$$W(\hat{r}) = b(1 - \hat{r}) + \left(\frac{b\hat{r}}{b\hat{r} + b(n - 2)r} \right) b(n - 1)(1 - r).$$

Taking the derivative with respect to $\hat{r}$ yields

$$\frac{\partial W(\hat{r})}{\partial \hat{r}} = -b + b(n - 1)(1 - r) \left(\frac{\hat{r} + b(n - 2)r - \hat{r}}{(\hat{r} + b(n - 2)r)^2} \right).$$

Setting $r = \hat{r} = r^\star$ yields

$$\left. \frac{\partial W(\hat{r})}{\partial \hat{r}} \right|_{r=\hat{r}} = -b + b(n - 1)(1 - r^\star) \left(\frac{n - 2}{(n - 1)^2 r^\star} \right).$$

Set $r = \hat{r} = r^\star$ and set the entire expression equal to zero:

$$0 = -b + b(n-1)(1-r^\star) \left(\frac{n - 2}{(n - 1)^2 r^\star} \right) = -1 + (1 - r^\star) \left(\frac{n - 2}{(n - 1)r^\star} \right)$$

Solving for $r^\star$ yields the desired result. The ESS sex ratio is biased in small groups even without sib mating, therefore mating between sibs is not the cause of the biased sex ratio. ∎

3. Cooperation among siblings

On the boundary R is a constant. Thus

$$m'f' = \frac{m^2}{R} a(R - m).$$

Taking the derivative with respect to m yields

$$\frac{d}{dm} m'f' = a(2m - 3m^2/R).$$

Set this expression equal to zero:

$$0 = 2R - 3m.$$

Solving for m yields the desired result. ∎

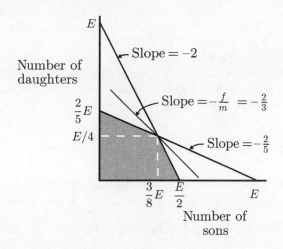

Figure E.6: The fitness set for Chapter 7, Problem 4, part a.

4. More than one constraint

(a) As is shown in Figure E.6, these two constraints lead to a polygonal fitness set. To find the values of m and f at the interior vertex, set

$$2m + f = m + \frac{5}{2}f \Rightarrow m = \frac{3}{2}f,$$

and thus

$$E = 3f + f \Rightarrow f = \frac{E}{4}$$

and

$$m = \frac{3}{2}\left(\frac{E}{4}\right) = \frac{3E}{8}.$$

(b) To find the ESS sex ratio we use the Shaw-Mohler theorem. The vertex is an obvious spot to try because there are a range of slopes which would produce the ESS sex ratio there. At the vertex

$$\frac{f_0}{m_0} = \frac{\frac{E}{4}}{\frac{3E}{8}} = \frac{2}{3}.$$

Thus, the slope of the line $f - f_0 = -\left(\frac{f_0}{m_0}\right)(m - m_0)$ at the vertex is $-2/3$, which is less than the slope of the boundary to the left and greater than the slope on the right, and therefore 2/3 is the ESS ratio of females to males. No other ratio can be an ESS because the slope associated with points to the left will be even less than $-2/3$ and therefore will intersect the boundary of the fitness set. Points to the right will be greater than $-2/3$ and therefore will also intersect the fitness set. ∎

Chapter 8

1. Correlated response to selection and evolutionary equilibria

From the chapter we know that at equilibrium

$$0 = G_x \left. \frac{\partial \ln W}{\partial x} \right|_{\bar{x},\bar{y}} + B_{xy} \left. \frac{\partial \ln W}{\partial y} \right|_{\bar{x},\bar{y}}$$

$$0 = B_{xy} \left. \frac{\partial \ln W}{\partial x} \right|_{\bar{x},\bar{y}} + G_y \left. \frac{\partial \ln W}{\partial y} \right|_{\bar{x},\bar{y}}.$$

This is a system of homogeneous linear equations, and so so has a nontrivial solution only if the determinant of the matrix of constants is zero. But the determinant is $G_x G_y - B_{xy}^2$. This equals zero only if

$$\frac{B_{xy}}{(G_x G_y)^{\frac{1}{2}}} = 1.$$

The left-hand side is the correlation. If this is not satisfied, the only solution is that both derivatives of the logarithm of fitness have to be zero. The reason is that, unless the correlation is one, there is some independent genetic variation in both traits. Thus unless both selection gradients are zero, the trait can't be in equilibrium. ∎

2. Runaway with biased mutation

(a) From the chapter, we know that the combination of a costly male trait and a costly female choice leads to the expressions

$$\Delta \bar{p} = \frac{1}{2} B_{tp}(a\bar{p} - 2c\bar{t}) - G_p b\bar{p}$$

$$\Delta \bar{t} = \frac{1}{2} G_t(a\bar{p} - 2c\bar{t}) - B_{tp} b\bar{p}.$$

Since mutation only directly affects $\bar{t}$, and the mean is reduced a constant amount μ, subtract μ from the right side of $\Delta\bar{t}$ to get the desired recursions.

(b) Set $\Delta\bar{p} = \Delta\bar{t} = 0$ and solve for the equilibrium values of the female preference and the male trait, $\hat{p}$ and $\hat{t}$, respectively:

$$\hat{p}(aB_{tp}(a - 2c\hat{t}) - 2G_p b\hat{p} = 0$$
$$G_t(a\hat{p} - 2c\hat{t}) - B_{tp}b\hat{p} = 2\mu.$$

Rearranged, these give

$$\hat{p}(aB_{tp} - 2bG_p) - 2cB_{tp}\hat{t} = 0$$
$$\hat{p}(aG_t - 2bB_{tp}) - 2cG_t\hat{t} = 2\mu.$$

The top equation becomes

$$\hat{p} = \frac{2cB_{tp}\hat{t}}{aB_{tp} - 2bG_p}.$$

Substituting this expression for $\hat{t}$ into the second equation yields

$$\frac{2cB_{tp}\hat{t}}{aB_{tp} - 2bG_p}(aG_t - 2bB_{tp}) - 2cG_t\hat{t} = 2\mu$$

$$\hat{t}\left(\frac{cB_{tp}(aG_t - 2bB_{tp}) - cG_t(aB_{tp} - 2bG_p)}{aB_{tp} - 2bG_p}\right) = \mu$$

$$\hat{t} = \frac{\mu(aB_{tp} - 2bG_p)}{2cb\left(G_tG_p - B_{tp}^2\right)}.$$

Substitute this back into the expression for $\hat{p}$:

$$\hat{p} = \frac{2cB_{tp}}{aB_{tp} - 2bG_p}\left(\frac{\mu(B_{tp} - 2bG_p)}{2cb\left(G_tG_p - B_{tp}^2\right)}\right)$$

$$= \frac{\mu B_{tp}}{b\left(G_tG_p - B_{tp}^2\right)}.$$

(c) The ratio of the means is

$$\frac{\hat{p}}{\hat{t}} = \frac{\mu B_{tp}}{b\left(G_tG_p - B_{tp}^2\right)}\left(\frac{2cb\left(G_tG_p - B_{tp}^2\right)}{\mu(aB_{tp} - 2bG_p)}\right) = \frac{2cB_{tp}}{aB_{tp} - 2bG_p}.$$

When b is small this becomes

$$\frac{\hat{p}}{\hat{t}} \approx \frac{2c}{a},$$

which is the expression for the line of equilibrium in the no-mutation case.

(d)   The expressions for the mean are both proportional to μ/b, so if the costs of female choice are small, the means can be displaced a long way from zero. ∎

3. When selection limits the runaway process

(a)   The equations for the changes in the means are given by

$$\Delta\bar{p} = \frac{1}{2}B_{tp} \left.\frac{\partial \ln W}{\partial t}\right|_{\bar{t},\bar{p}} = \frac{1}{2}B_{tp}(a\bar{p} - 4c\bar{t}^3) - G_p b\bar{p}$$

$$\Delta\bar{t} = \frac{1}{2}G_t \left.\frac{\partial \ln W}{\partial t}\right|_{\bar{t},\bar{p}} = \frac{1}{2}G_t(a\bar{p} - 4c\bar{t}^3) - B_{tp} b\bar{p}.$$

(b)   Setting $b = 0$ simplifies the equations to

$$\Delta\bar{p} = \frac{1}{2}B_{tp}(a\bar{p} - 4c\bar{t}^3)$$

$$\Delta\bar{t} = \frac{1}{2}G_t(a\bar{p} - 4c\bar{t}^3).$$

The changes in the mean values of t and p will be zero if $a\bar{p} - 4c\bar{t}^3 = 0$. Thus there is a line of equilibrium that satisfies

$$\hat{p} = \frac{4c}{a}\bar{t}^3.$$

(c)   Dividing the expression for $\Delta\bar{p}$ by the expression for $\Delta\bar{t}$ yields

$$\frac{\Delta\bar{p}}{\Delta\bar{t}} = \frac{B_{tp}}{G_t}.$$

(d)   Plotting the line of equilibrium and the phase space trajectories yields the plot shown in Figure E.7.

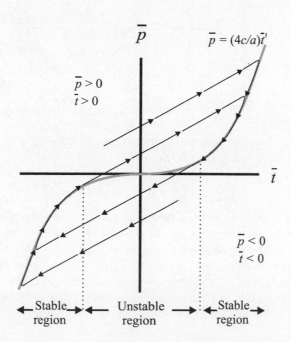

Figure E.7: The line of equilibrium and phase space trajectories. The grey curve is the line of equilibrium when female choice is costless. Away from this line, the population moves along the straight lines with slope B_{tp}/G_t in the direction given by the arrows. With costless female choice the line is stable when its slope is greater than B_{tp}/G_t and unstable when its slope is less than this ratio. When choice is costly, populations move along stable segments of the line toward $\bar{p} = 0$, the value that minimizes the cost of female choice. However, eventually the line becomes unstable and the population runs away. This runaway is eventually limited by increasing male mortality, and the cycle continues.

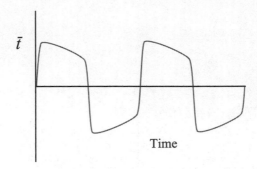

Figure E.8: The trajectory of $\bar{t}$ plotted against time. The trajectory for $\bar{p}$ is similar but shifted to be out of phase.

(e)   The trajectories of the mean trait values are sketched in Figure E.7. When the slope of the line of equilibrium is less than the slope of the phase space trajectories, the line is stable and the means move slowly along it toward $(0,0)$. However, as the slope of the line lessens it becomes unstable and the population moves rapidly to the line in the opposite quadrant. This leads to oscillations like those sketched in Figure E.8. ■

Bibliography

Alexander, R. D. (1987). *The Biology of Moral Systems*. Aldine De Gruyter, New York.

Alexander, R. D. and Sherman, P. W. (1977). Local mate competition and parental investment in social insects. *Science*, 196:494–500.

Andersson, M. (1994). *Sexual Selection*. Princeton University Press, Princeton, NJ.

Aoki, K. (1982). A condition for group selection to prevail over counteracting individual selection. *Evolution*, 36:432–442.

Aviles, L. (1993). Interdemic selection and the sex ratio: a social spider perspective. *The American Naturalist*, 142:320–345.

Aviles, L. (2002). Solving the freeloaders paradox: genetic associations and frequency dependent selection in the evolution of cooperation among nonrelatives. *Proceedings of the National Academy of Sciences, USA*, 99:14268–14273.

Axelrod, R. M. (1984). *The Evolution of Cooperation*. Basic Books, New York.

Axelrod, R. M. (1986). An evolutionary approach to norms. *American Political Science Review*, 80:1095–1111.

Axelrod, R. M. and Dion, D. (1988). The further evolution of cooperation. *Science*, 242:1385–1390.

Axelrod, R. M. and Hamilton, W. (1981). The evolution of cooperation in biological systems. *Science*, 211:1390–1396.

Axelrod, R. M., Hammond, R. A., and Grafen, A. (2004). Altruism via kin-selection strategies that rely on arbitrary tags with which they coevolve. *Evolution*, 58:1833–1838.

Barton, N. and Turelli, M. (1989). Evolutionary quantitative genetics: how little do we know? *Annual Review of Genetics*, 23:337–370.

Barton, N. and Turelli, M. (1991). Natural and sexual selection on many loci. *Genetics*, 127:229–255.

Barton, N. H. and Keightley, P. D. (2002). Understanding quantitative genetic variation. *Nature Reviews Genetics*, 3:11–21.

Batali, J. and Kitcher, P. (1995). Evolution of altruism in optional and compulsory games. *Journal of Theoretical Biology*, 175:161–171.

Bergstrom, C. T. and Lachmann, M. (1997). Signalling among relatives. I. When is signalling too costly? *Philosophical Transactions of the Royal Society of London, Series B*, 352:609–617.

Bergstrom, C. T. and Lachmann, M. (1998). Signalling among relatives. III. Talk is cheap. *Proceedings of the National Academy of Sciences, USA*, 95:5100–5105.

Bergstrom, T. (2002). Evolution of social behavior: Individual and group selection. *Journal of Economic Perspectives*, 16:67–88.

Bishop, D. and Cannings, C. (1978). A generalized war of attrition. *Journal of Theoretical Biology*, 70:85–125.

Boerlijst, M. C., Nowak, M. A., and Sigmund, K. (1997). The logic of contrition. *Journal of Theoretical Biology*, 185:281–293.

Boyd, R. (1982). Density-dependent mortality and the evolution of social interactions. *Animal Behaviour*, 30:972–982.

Boyd, R. (1989). Mistakes allow evolutionary stability in the repeated prisoner's dilemma. *Journal of Theoretical Biology*, 136:47–56.

Boyd, R., Gintis, H., Bowles, S., and Richerson, P. J. (2003). The evolution of altruistic punishment. *Proceedings of the National Academy of Sciences, USA*, 100:3531–3535.

Boyd, R. and Lorberbaum, J. P. (1987). No pure strategy is evolutionarily stable in the repeated prisoner's dilemma game. *Nature*, 327:58–59.

Boyd, R. and Richerson, P. J. (1985). *Culture and the Evolutionary Process*. University of Chicago Press, Chicago.

Boyd, R. and Richerson, P. J. (1988). The evolution of reciprocity in sizeable groups. *Journal of Theoretical Biology*, 132:337–356.

Boyd, R. and Richerson, P. J. (1989). The evolution of indirect reciprocity. *Social Networks*, 11:213–236.

Boyd, R. and Richerson, P. J. (1990). Group selection among alternative evolutionarily stable strategies. *Journal of Theoretical Biology*, 145:331–342.

Boyd, R. and Richerson, P. J. (1992). Punishment allows the evolution of cooperation (or anything else) in sizable groups. *Ethology and Sociobiology*, 13:171–195.

Boyd, R. and Richerson, P. J. (1995). Why does culture increase human adaptability? *Ethology and Sociobiology*, 16:125–143.

Boyd, R. and Richerson, P. J. (2002). Group beneficial norms spread rapidly in a structured population. *Journal of Theoretical Biology*, 215:287–296.

Brandon, R. N. (1990). *Adaptation and Environment*. Princeton University Press, Princeton, NJ.

Brilot, B. and Johnstone, R. A. (2002). The limits to cost-free signalling of need between relatives. *Proceedings of the Royal Society, Series B*, 270:1055–1060.

Brown, G. R. and Silk, J. B. (2002). Reconsidering the null hypothesis: is maternal rank associated with birth sex ratios in primate groups? *Proceedings of the National Academy of Sciences, USA*, 99:11252–11255.

Bulmer, M. G. (1985). *The Mathematical Theory of Quantitative Genetics*. Oxford University Press, London.

Bulmer, M. G. and Taylor, P. D. (1980). Dispersal and the sex ratio. *Nature*, 284:448–449.

Camerer, C. (2003). *Behavioral Game Theory, Experiments in Strategic Interaction*. Princeton University Press, Princeton, NJ.

Caswell, H. (2001). *Matrix Population Models: Construction, Analysis, Interpretation*. Sinauer Associates, Sunderland, MA.

Charlesworth, B. (1978). Some models of the evolution of altruistic behaviour between siblings. *Journal of Theoretical Biology*, 72:297–319.

Charnov, E. L. (1982). *The Theory of Sex Allocation*. Princeton University Press, Princeton, NJ.

Clutton-Brock, T. (2002). Breeding together: kin selection and mutualism in cooperative vertebrates. *Science*, 296:69–72.

Clutton-Brock, T. H. and Parker, G. A. (1995). Punishment in animal societies. *Nature*, 373:209–216.

Coyne, J., Barton, N., and Turelli, M. (1997). A critique of Sewall Wright's shifting balance theory of evolution. *Evolution*, 51:643–671.

Crowley, P. H. (2000). Hawks, doves, and mixed-symmetry games. *Journal of Theoretical Biology*, 204:543–563.

Crowley, P. H. (2003). Origins of behavioural variability: categorical and discriminative assessment in serial contests. *Animal Behaviour*, 66:427–440.

Dawkins, R. (1976). *The Selfish Gene*. Oxford University Press, Oxford.

Dawkins, R. (1979). 12 misunderstandings of kin selection. *Zeitschrift für Tierpsychologie*, 51:184–200.

Dawkins, R. and Krebs, J. R. (1978). Animal signals: information or manipulation? In Krebs, J. and Davies, N., editors, *Behavioural Ecology: An Evolutionary Approach*, pages 282–309. Blackwell, Oxford.

Dugatkin, L. A. (1990). N-person games and the evolution of cooperation: a model based on predator inspection in fish. *Journal of Theoretical Biology*, 142:123–135.

Dugatkin, L. A. and Reeve, H. K. (1994). Behavioral ecology and the "levels of selection": dissolving the group selection controversy. *Advances in the Study of Behavior*, 23:101–133.

Dugatkin, L. A. and Wilson, D. S. (1991). Rover: A strategy for exploiting cooperators in a patchy environment. *The American Naturalist*, 138:687–701.

Endler, J. (1986). *Natural Selection in the Wild*. Princeton University Press, Princeton, NJ.

Enquist, M. and Leimar, O. (1983). Evolution of fighting behavior: the effect of variation in resource value. *Journal of Theoretical Biology*, 127:187–205.

Enquist, M. and Leimar, O. (1993). The evolution of cooperation in mobile organisms. *Animal Behaviour*, 45:747–757.

Eshel, I. and Feldman, M. W. (1984). Initial increase of new mutants and some continuity properties of ESS in two-locus systems. *The American Naturalist*, 124:631–640.

Fehr, E. and Gächter, S. (2002). Altruistic punishment in humans. *Nature*, 415:137–140.

Fisher, R. A. (1930, 1958). *The genetical theory of natural selection*. Dover, New York.

Frank, S. A. (1995). George Price's contributions to evolutionary genetics. *Journal of Theoretical Biology*, 175:373–388.

Frank, S. A. (1998). *Foundations of Social Evolution*. Princeton University Press, Princeton, NJ.

Gadagkar, R. (2001). *The Social Biology of* Ropalidia marginata: *Toward Understanding the Evolution of Eusociality*. Harvard University Press, Cambridge, MA.

Gardner, A. and West, S. A. (2004). Spite and the scale of competition. *Journal of Evolutionary Biology*, 17:1195–1203.

Gintis, H. (2000). *Game Theory Evolving*. Princeton University Press, Princeton, NJ.

Giraldeau, L.-A. and Caraco, T. (2000). *Social Foraging Theory*. Princeton University Press, Princeton, NJ.

Godfray, H. J. C. (1991). Signalling of need by offspring to their parents. *Nature*, 352:328–330.

Godfray, H. J. C. and Johnstone, R. A. (2000). Begging and bleating: the evolution of parent-offspring signalling. *Philosophical Transactions of the Royal Society of London, Series B*, 355:1581–1591.

Grafen, A. (1984). Natural selection, kin selection and group selection. In Krebs, J. and Davies, N., editors, *Behavioural Ecology: An Evolutionary Approach*, pages 62–84. Blackwell Scientific, Oxford.

Grafen, A. (1985). A geometric view of relatedness. *Oxford Surveys in Evolutionary Biology*, 2:28–90.

Grafen, A. (1987). The logic of divisively asymmetric contests: respect for ownership and the desperado effect. *Animal Behaviour*, 35:462–467.

Grafen, A. (1990a). Biological signals as handicaps. *Journal of Theoretical Biology*, 144:517–546.

Grafen, A. (1990b). Do animals really recognize kin? *Animal Behaviour*, 39:42–54.

Grafen, A. (1990c). Sexual selection unhandicapped by the Fisher process. *Journal of Theoretical Biology*, 144:473–516.

Grafen, A. (1998). Green beard as death warrant. *Nature*, 394:521–522.

Grafen, A. (2005). A theory of Fisher's reproductive value. In press, *Journal of Mathematical Biology*.

Greenwood, P. J. (1980). Mating systems, philopatry and dispersal in birds and mammals. *Animal Behaviour*, 28:1140–1162.

Haldane, J. B. S. (1955). Population genetics. *New Biology*, 18:34–51.

Hamilton, W. D. (1963). The evolution of altruistic behaviour. *The American Naturalist*, 97:354–356.

Hamilton, W. D. (1964). The genetical evolution of social behavior. *Journal of Theoretical Biology*, 7:1–52.

Hamilton, W. D. (1967). Extraordinary sex ratios. *Science*, 156:477–488.

Hamilton, W. D. (1970). Selfish and spiteful behaviour in an evolutionary model. *Nature*, 228:1218–1220.

Hamilton, W. D. (1971). Selection of selfish and altruistic behaviour in some extreme models. In Eisenberg, J. F. and Dillon, W. S., editors, *Man and Beast: Comparative Social Behavior*, pages 57–91. Smithsonian Press, Washington, DC.

Hamilton, W. D. (1975). Innate social aptitudes of man: an approach from evolutionary genetics. In Fox, R., editor, *Biosocial Anthropology*, pages 133–153. Malaby Press, London.

Hamilton, W. D. (1996). *Narrow Roads of Gene Land: The Collected Papers of W. D. Hamilton: Evolution of Social Behaviour*. Oxford University Press, Oxford.

Hamilton, W. D. and May, R. M. (1977). Dispersal in stable habitats. *Nature*, 269:578–581.

Hammerstein, P. (1981). The role of asymmetries in animal contests. *Animal Behaviour*, 29:193–205.

Hammerstein, P. (1996). Darwinian adaptation, population genetics and the streetcar theory of evolution. *Journal of Mathematical Biology*, 34:511–530.

Hardy, I. C. W., editor (2000). *Sex Ratios: Concepts and Research Methods*. Cambridge University Press, Cambridge.

Hastings, A. (1997). *Population Biology: Concepts and Models*. Springer-Verlag, Berlin.

Henrich, J. and Boyd, R. (2001). Why people punish defectors— weak conformist transmission can stabilize costly enforcement of norms in cooperative dilemmas. *Journal of Theoretical Biology*, 208:79–89.

Hilborn, R. and Mangel, M. (1997). *The Ecological Detective: Confronting Models with Data*. Princeton University Press, Princeton, NJ.

Hirshleifer, J. (1977). Economics from a biological point of view. *Journal of Law and Economics*, 20:1–52.

Houston, A. I. and McNamara, J. M. (1988). Fighting for food: a dynamic version of the hawk-dove game. *Evolutionary Ecology*, 2:51–64.

Houston, A. I. and McNamara, J. M. (1999). *Models of Adaptive Behaviour: An Approach Based on State*. Cambridge University Press, Cambridge.

Hurd, P. L. and Enquist, M. (2005). A strategic taxonomy of biological communication. *Animal Behaviour*, 70:1155–1170.

Huxley, J. S. (1938a). Darwin's theory of sexual selection and the data subsumed by it in the light of recent research. *The American Naturalist*, 72:416–433.

Huxley, J. S. (1938b). The present standing of the theory of sexual selection. In de Beer, G. R., editor, *Evolution: Essays on Aspects of Evolutionary Biology*, pages 11–42. Clarendon Press, Oxford.

Iwasa, Y., Pomiankowski, A., and Nee, S. (1991). The evolution of costly mate preferences II. The "handicap" principle. *Evolution*, 45:1431–1442.

Johnstone, R. A. (2000). Models of reproductive skew: a review and synthesis. *Ethology*, 106:5–26.

Johnstone, R. A. (2002). Signalling of need, sibling competition, and the cost of signalling. *Proceedings of the National Academy of Sciences, USA*, 96:12644–12649.

Johnstone, R. A. and Grafen, A. (1992). The continuous Sir Philip Sidney game: a simple model of biological signalling. *Journal of Theoretical Biology*, 156:215–234.

Joshi, N. V. (1987). Evolution of cooperation by reciprocation within structured demes. *Journal of Genetics*, 66:69–84.

Kimura, M. and Crow, J. (1964). The number of alleles that can be maintained in a finite population. *Genetics*, 49:725–738.

Kirkpatrick, M. (1982). Sexual selection and the evolution of female choice. *Evolution*, 3:1–12.

Kokko, H. (2003). Are reproductive skew models evolutionarily stable? *Proceedings of the Royal Society of London, Series B*, 270:265–270.

Kokko, H., Brooks, R., Jennions, M. D., and Morely, J. (2003). The evolution of mate choice and mating biases. *Proceedings of the Royal Society of London, Series B*, 270:653–664.

Kokko, H., Brooks, R., McNamara, J. M., and Houston, A. I. (2002). The sexual selection continuum. *Proceedings of the Royal Society of London, Series B*, 269:1331–1340.

Lachmann, M. and Bergstrom, C. T. (1998). Signalling among relatives. II. Beyond the Tower of Babel. *Theoretical Population Biology*, 54:146–160.

Lande, R. (1981). Models of speciation by sexual selection on polygenic traits. *Proceedings of the National Academy of Sciences, USA*, 78:3721–3725.

Leimar, O. (1996). Life history analysis of the Trivers-Willard sex-ratio problem. *Behavioral Ecology*, 7:316–325.

Leimar, O. (1997). Repeated games: a state space approach. *Journal of Theoretical Biology*, 184:471–498.

Leimar, O. and Hammerstein, P. (2001). Evolution of cooperation through indirect reciprocity. *Proceedings of the Royal Society of London, Series B*, 268:745–753.

Maynard Smith, J. (1976). Sexual selection and the handicap principle. *Journal of Theoretical Biology*, 57:239–242.

Maynard Smith, J. (1982). *Evolution and the Theory of Games*. Cambridge University Press, Cambridge.

Maynard Smith, J. (1991). Honest signalling: the Philip Sidney game. *Animal Behaviour*, 42:1034–1035.

Maynard Smith, J. and Harper, D. (2003). *Animal Signals*. Oxford University Press, Oxford.

Maynard Smith, J. and Parker, G. A. (1976). The logic of asymmetric contests. *Animal Behaviour*, 24:159–175.

Maynard Smith, J. and Price, G. R. (1973). The logic of animal conflict. *Nature*, 146:15–18.

McNamara, J. M. and Houston, A. I. (1996). State-dependent life histories. *Nature*, 380:215–221.

Mead, L. and Arnold, S. (2004). Quantitative genetic models of sexual selection. *Trends in Ecology and Systematics*, 19:264–271.

Michod, R. E. (1979). Genetical aspects of kin selection: effects of inbreeding. *Journal of Theoretical Biology*, 81:223–233.

Michod, R. E. (1980). Evolution of interactions in family-structured populations: mixed mating models. *Genetics*, 96:275–296.

Milinski, M. (1987). Tit for tat in sticklebacks and the evolution of cooperation. *Nature*, 325:433–437.

Mock, D. W. and Parker, G. A. (1998). *The Evolution of Sibling Rivalry*. Oxford University Press, Oxford.

Nowak, M. A. and Sigmund, K. (1993). A strategy of win-stay, lose-shift that outperforms tit-for-tat in the prisoner's dilemma game. *Nature*, 364:56–58.

Nowak, M. A. and Sigmund, K. (1998a). The dynamics of indirect reciprocity. *Journal of Theoretical Biology*, 194:561–574.

Nowak, M. A. and Sigmund, K. (1998b). Evolution of indirect reciprocity by image scoring. *Nature*, 393:573–577.

Nowak, M. A. and Sigmund, K. (2005). The evolution of indirect reciprocity. *Nature*, 437:1291–1298.

Panchanathan, K. and Boyd, R. (2003). A tale of two defectors: the importance of standing for the evolution of indirect reciprocity. *Journal of Theoretical Biology*, 224:115–126.

Panchanathan, K. and Boyd, R. (2004). Indirect reciprocity can stabilize cooperation without the second-order free rider problem. *Nature*, 432:499–502.

Parker, G. A. (1974). Assessment strategy and the evolution of fighting behaviour. *Journal of Theoretical Biology*, 47:223–243.

Parker, G. A. and Thompson, E. A. (1980). Dung fly struggles: a test of the war of attrition. *Behavioral Ecology and Sociobiology*, 7:37–44.

Peck, S. L., Ellner, S. P., and Gould, F. (1998). A spatially explicit stochastic model demonstrates the feasibility of Wright's shifting balance theory. *Evolution*, 52:1834–1839.

Pen, I. and Weissing, F. J. (2002). Optimal sex allocation: steps towards a mechanistic theory. In Hardy, I. C. W., editor, *Sex Ratios: Concepts and Research Methods*, pages 26–47. Cambridge University Press, Cambridge.

Pennisi, E. (2005). How did cooperative behavior evolve? *Science*, 309:75.

Perrin, N. and Goudet, J. (2001). Inbreeding, kinship, and the evolution of natal dispersal. In Clobert, J., Danchin, E., Dhont, A. A., and Nichols, J. D., editors, *Dispersal*, pages 123–142. Oxford University Press, Oxford.

Pomiankowski, A. (1987). The cost of choice in sexual selection. *Journal of Theoretical Biology*, 128:195–218.

Pomiankowski, A. and Iwasa, Y. (1993). Evolution of multiple sexual preferences by Fisher runaway process of sexual selection. *Proceedings of the Royal Society of London, Series B*, 253:173–181.

Price, G. R. (1970). Selection and covariance. *Nature*, 227:520–521.

Price, G. R. (1972). Extension of covariance selection mathematics. *Annals of Human Genetics*, 35:485–490.

Queller, D. C. (1994). Genetic relatedness in viscous populations. *Evolutionary Ecology*, 8:70–73.

Queller, D. C. (1996). The measurement and meaning of inclusive fitness. *Animal Behaviour*, 51:229–232.

Reeve, H. K. (1998). Game theory, reproductive skew, and nepotism. In Dugatkin, L. and Reeve, H., editors, *Game Theory and Animal Behavior*, pages 118–145. Oxford University Press, Oxford.

Rice, S. (2004). *Evolutionary Theory: Mathematical and Conceptual Foundations*. Sinauer, Sunderland, MA.

Richerson, P. J. and Boyd, R. (2005). *Not by Genes Alone: How Culture Transformed Human Evolution.* University of Chicago Press, Chicago.

Riechert, S. E. (1984). Games spiders play 3. Cues underlying context associated changes in agonistic behavior. *Animal Behaviour*, 32:1–15.

Riechert, S. E. (1986). Spider fights as a test of evolutionary game theory. *American Scientist*, 74:604–610.

Riechert, S. E. and Hammerstein, P. (1983). Game theory in the ecological context. *Annual Review of Ecology and Systematics*, 14:377–409.

Rogers, A. R. (1988). Does biology constrain culture? *American Anthropologist*, 90:819–831.

Rogers, A. R. (1990). Group selection by selective emigration: the effects of migration and kin structure. *American Naturalist*, 135:398–413.

Roughgarden, J. (1979, 1996). *Theory of Population Genetics and Evolutionary Ecology.* Prentice Hall, Upper Saddle River, NJ.

Roze, D. and Rousset, F. (2004). The robustness of Hamilton's rule with inbreeding and dominance: kin selection and fixation probabilities under partial sib mating. *The American Naturalist*, 164:214–231.

Ryan, M. J. (1998). Receiver biases, sexual selection and the evolution of sex differences. *Science*, 281:1999–2003.

Selten, R. (1975). Reexamination of the perfectness concept for equilibrium points in extensive games. *International Journal of Game Theory*, 4:25–55.

Selten, R. and Hammerstein, P. (1984). Gaps in Harley's argument on the evolutionary stable learning rules and in the logic of "tit for tat." *The Behavioral and Brain Sciences*, 7:115–116.

Seyfarth, R. and Cheney, D. (1984). Grooming, alliances and reciprocal altruism in vervet monkeys. *Nature*, 308:541–543.

Shaw, R. F. and Mohler, J. D. (1953). The selective advantage of the sex ratio. *The American Naturalist*, 87:337–342.

Sherman, P. W. (1977). Nepotism and the evolution of alarm calls. *Science*, 197:1246–1253.

Shuster, S. M. and Wade, M. J. (2003). *Mating Systems and Strategies*. Princeton University Press, Princeton, NJ.

Silk, J. B., Kaldor, E., and Boyd, R. (2000). Cheap talk when interests conflict. *Animal Behaviour*, 59:423–432.

Sober, E. and Wilson, D. S. (1998). *Unto Others: The Evolution and Psychology of Unselfish Behavior*. Harvard University Press, Cambridge, MA.

Spence, A. M. (1973). Job market signaling. *Quarterly Journal of Economics*, 83:355–377.

Sugden, R. (1986). *The Economics of Rights, Co-operation and Welfare*. Blackwell Publishers, London.

Taylor, M. (1976). *Anarchy and Cooperation*. John Wiley and Sons, New York.

Taylor, P. D. (1990). Allele-frequency change in a class-structured population. *The American Naturalist*, 135:95–106.

Taylor, P. D. (1992). Altruism in viscous populations—an inclusive fitness approach. *Evolutionary Ecology*, 6:352–356.

Taylor, P. D. (1996). Inclusive fitness arguments in genetic models of behavior. *Journal of Mathematical Biology*, 34:654–674.

Taylor, P. D. and Bulmer, M. G. (1980). Local mate competition and the sex ratio. *Journal of Theoretical Biology*, 86:409–419.

Taylor, P. D. and Frank, S. A. (1996). How to make a kin selection model. *Journal of Theoretical Biology*, 180:27–37.

Tinbergen, N. (1952). "Derived" activities; their origin, causation, biological significance, origin, and emancipation during evolution. *The Quarterly Review of Biology*, 27:1–32.

Trivers, R. L. (1971). The evolution of reciprocal altruism. *Quarterly Journal of Biology*, 46:35–57.

Trivers, R. L. (1974). Parent-offspring conflict. *American Zoologist*, 14:249–264.

Trivers, R. L. and Willard, D. E. (1973). Natural selection of parental ability to vary the sex ratio of offspring. *Science*, 179:90–92.

Turchin, P. (2003). *Historical Dynamics*. Princeton University Press, Princeton, NJ.

Uyenoyama, M. K. (1984). Inbreeding and the evolution of altruism under kin selection: effects on relatedness and group structure. *Evolution*, 48:778–779.

Veblen, T. (1899, 1992). *The Theory of the Leisure Class: An Economic Study of Institutions*. Transaction Publishers, New Brunswick, NJ.

Vehrencamp, S. L. (1979). The roles of individual, kin and group selection in the evolution of sociality. In Marler, P. and Vandenbergh, J., editors, *Handbook of Behavioural Neurobiology 3. Social Behaviour and Communication*, pages 240–283. Plenum Press, New York.

Wade, M. J. (1978). A critical review of the models of group selection. *Quarterly Review of Biology*, 53:101–114.

Wade, M. J. (1985). Soft selection, hard selection, kin selection, and group selection. *The American Naturalist*, 125:61–73.

Wade, M. J. and Goodnight, C. J. (1998). The theories of Fisher and Wright in the context of metapopulations: when nature does many small experiments. *Evolution*, 52:1537–1553.

Wallace, B. (1968). Polymorphism, population size, and genetic load. In Lewontin, R., editor, *Population Biology and Evolution*, pages 87–108. Syracuse University Press, Syracuse, New York.

Washburn, S. L. (1978). Human behavior and the behavior of other animals. *American Psychologist*, 33:405–418.

Weatherhead, P. and Robinson, R. J. (1979). Offspring quality and the polygyny threshold: "the sexy son hypothesis." *The American Naturalist*, 113:201–208.

West, S. A. and Sheldon, B. C. (2002). Constraints in the evolution of sex ratio adjustment. *Science*, 295:1685–1688.

Wilkinson, R. (1984). Reciprocal food sharing in the vampire bat. *Nature*, 308:181–184.

Wilson, D. S., Pollock, G., and Dugatkin, L. (1992). Can altruism evolve in purely viscous populations? *Evolutionary Ecology*, 6:331–341.

Wilson, E. O. (1975). *Sociobiology: The New Synthesis*. Belknap Press, Cambridge, MA.

Wimsatt, W. C. (1987). False models as means to truer theories. In Nitecki, M. and Hoffman, H., editors, *Neutral Models in Biology*, pages 23–55. Oxford University Press, Oxford.

Wright, S. (1922). Coeficients of inbreeding and relationship. *The American Naturalist*, 56:330–338.

Wright, S. (1931). Evolution in mendelian populations. *Genetics*, 16:97–159.

Wynne-Edwards, V. C. (1962). *Animal Dispersion in Relation to Social Behaviour*. Oliver and Boyd, Edinburgh.

Zahavi, A. (1975). Mate selection—a selection for a handicap. *Journal of Theoretical Biology*, 53:205–213.

Index